Technology and the African-American Experience

Technology and the African-American Experience
Needs and Opportunities for Study

edited by Bruce Sinclair

published in cooperation with the Lemelson Center for the Study of Invention and Innovation at the Smithsonian Institution

The MIT Press
Cambridge, Massachusetts
London, England

Set in Sabon by The MIT Press. Printed and bound in the United States of America.

Library of Congress Cataloging-in-Publication Data

Technology and the African–American experience / edited by Bruce Sinclair.
p. cm.
Includes bibliographical references and index.
ISBN 0-262-19504-6 (alk. paper)
1. Technology—United States—History. 2. Technology—Social aspects—United States—History. 3. African Americans—History. 4. United States—Race relations. I. Sinclair, Bruce.
T21.T4295 2004
608.996'073--dc22   2003066636

10 9 8 7 6 5 4 3 2 1

Contents

Museums and the Interpretation of African-American History   187
Lonnie Bunch

A Bibliography of Technology and the African-American Experience   197
Amy Sue Bix

Preface

For the historians of American technology of my generation, Brooke Hindle's *Early American Technology*: *Needs and Opportunities for Study* (Institute of Early American History and Culture, 1966) promised what we all wanted to believe—that our subject would prove central to the understanding of American history, and that there was a crucial need for research and teaching in it. Looking for ideas and bibliography, we turned those pages time and again, and as an academic I grew up with the idea that a book like that was a natural and useful historiographical form. To make it easier to explain what kind of enterprise it was to be, I gave this volume the same subtitle Hindle used. And my book's purpose, as is true of Hindle's, is really pretty simple; it aims to encourage people to explore the historical connections between race and technology. The hope here is that teachers will see useful ideas and material to include in their courses, that students will find the topic interesting to investigate, and that scholars will discover its research possibilities.

There are some differences, though, in these parallel excursions. My book has multiple authors, and I no longer believe that technology, studied from the inside out, yields the single best key to our past. On the other hand, quite a strong argument can be made for the proposition that race and technology are two of the most powerful themes in American history, that they have always been intertwined, that this relationship has yet to be understood, and that its analysis will lead us in highly rewarding new directions. The essays presented here mean to begin that inquiry. They are neither chronologically nor topically comprehensive, precisely because this is a subject matter yet to be developed. But even in their miscellaneous character, all of us connected with this project trust they may help reveal the promise that the study of race and technology holds.

As with all things, this book has a past. The idea of writing a history of technology that included people of color came out of a curriculum project masterminded by Susan Smulyan, on behalf of the Society for the History of Technology, with funding from the National Science Foundation. The main idea behind that project was to use the history of technology to encourage women and minority students in middle and high schools to think about careers in engineering and science. But to achieve that end we needed to reinvent our subject, to discover the hidden populations in it, and Susan proved to be wonderfully smart at that.

The next step came in 1994 with a project I directed that was jointly funded by the NEH and the NSF and administered with wit and wisdom by the Georgia Humanities Council. Called Technology and the African-American Experience (good titles get repeated), the project involved a workshop for teachers and scholars, where Smulyan, an Associate Professor in the Department of American Civilization at Brown University, Carroll Pursell, Chair of the History Department at Case Western Reserve University, Steve Lubar, Chair of the Division of the History of Technology, Smithsonian National Museum of American History, and Robert Hall, Department of African-American Studies at Northeastern University, presented position papers. Their insights have fundamentally informed everything since, and it is a treat to be indebted to one's best friends for such perceptive thinking. The workshop was followed by a series of public lectures in Atlanta, given by Linda McMurray of North Carolina State University, Cathleen Lewis and Lonnie Bunch of the Smithsonian Institution, and Warren Whatley of the University of Michigan. The open discussions held after the lectures were moderated with great effectiveness by Dr. Billie Gaines of Atlanta, and the strong audience participation made it clear that something more was needed. Happily, David Morton and Aristotelis Tympas, then graduate research assistants of mine at Georgia Tech, had done some splendid work on the subject. The materials they gathered, along with those position papers, suggested the book.

It was at this point that the Lemelson Center for the Study of Invention and Innovation, at the National Museum of American History, stepped in with the financial support necessary to bring authors and editors together—for both this volume and its companion, *Technology and the African-American Experience: A Documentary History*, which Carroll Pursell has edited. It has taken much longer to finish this job than any of us ever imagined, so besides the funding, I have as well to thank Art Molella and his staff at the Lemelson Center for their good-natured patience.

While at the Smithsonian, three bright, young scholars—Rayvon Fouche, Angela Lakewete, and Jill Snider—helped conceptualize both of the books, and I am grateful for their collaboration. Nikki Stanton, than a summer intern from Brown University and now a Ph.D. student at the University of Michigan, worked on the project, too—happily for me and for everyone else at the museum who met her. At that place I also learned a great deal in conversations with Alonzo Smith, Linda Tucker, Jim Horton, and Michelle Gates Moresi, as well as with John Vlach of George Washington University. Joe Trotter of Carnegie Mellon University was kind enough to offer helpful suggestions. I am particularly grateful to Charlie Haines, an independent scholar, to Michal McMahon of West Virginia University, and to Bill Shade of Lehigh University, all of whom read my introductory essay. I probably should have paid even more attention to their advice, though they cannot be blamed for that. Bill Scott, also of Lehigh University, has been a warm supporter of the project for a long time. But the biggest obligation is to Gail Cooper, who had the best ideas, the deepest interest, and the loving heart.

Technology and the African-American Experience

Integrating the Histories of Race and Technology

Bruce Sinclair

This volume brings together two subjects strongly connected but long segregated from each other. The history of race in America has been written as if technologies scarcely existed, and the history of technology as if it were utterly innocent of racial significance. Neither of these assumptions bears scrutiny. Indeed, in both cases the very opposite is true; an ancient and pervasive set of bonds links their histories. But there is little by way of an established literature that directly explores this relationship, nor a body of teaching that unites the two subjects. So we must begin the project of constructing a joint history by re-thinking our own assumptions, by borrowing useful ideas from related fields of scholarship, and by selecting examples of method and subject matter that promise fruitful lines of investigation—and in that fashion lay a groundwork. That is why this book is subtitled "Needs and Opportunities for Study." Its goal is simply to open up the topic for further exploration.

There are reasons why the past we seek to reveal has been so long denied, and racial prejudice dominates all of them. But more particularly, perceptions about inventiveness, presentations in our history about the nation-building role of technological talent, and the disciplinary boundaries between the fields of study themselves—as well as the politics that drove their own development—have all served to mask reality, and they are among the issues I want to consider in this introductory essay. A good place to start is with our oldest, most obvious attitudes. White Americans, including those as committed to Enlightenment ideals as Thomas Jefferson—even as he corresponded with Benjamin Banneker, the African-American astronomer and almanac maker— believed the black people among them were mentally inferior, and by that they didn't just mean a capacity for advanced intellectual accomplishment.[1] What good would freedom be, one Southern planter put it, to a field hand whose highest faculties were taxed "to discriminate between cotton and crop-grass, and to strike one with a hoe without hitting the other"?[2] Crude preconceptions of mental inferiority went well beyond simple tool using to include almost any aptitude for technological competence, and these notions flowered in the basic conditions of forced servitude. Owners linked the supposed endurance for hard, menial labor to brutish intelligence, and then justified enslavement on the grounds of such limited capacities. Besides that casual kind of rationalization, a substantial eighteenth-century literature invidiously compared African and other non-Western civilizations in terms of their relative backwardness in science and technology, making it easy for Europeans and Americans to take it as given that inventive talent was not to be found in any people of color.[3]

The idea that technical competence was related to race grew even more fixed with time. Even in the relatively tolerant city of Philadelphia, the Franklin Institute, established in 1824 explicitly to encourage the development of the mechanic arts, refused to allow blacks to attend lectures or classes. As Nina Lerman shows in her essay in this volume, the city's educational institutions increasingly planned occupations for its black students that required only minimal training. The great industrial expositions of the latter part of the nineteenth century made the same point graphically in the contrasts they drew between exhibits of the savagery of the dark-skinned peoples of the world and the brilliant flowering of civilized progress epitomized in Chicago's 1893 "White City."[4] But rather than simply the shell or emblem of racist thinking, defining African-Americans as technically incompetent and then—in a kind of double curse—denying them access to education, control over complex machinery, or the power of patent rights lay at the heart of the distinctions drawn between black and white people in this country. That formulation always served important political, economic, and social functions, and it is fundamentally why race and technology have for such a long time seemed different, even immiscible, categories of analysis. Racism may have colored all our history, but it whitened the national narrative.

Now, without looking very hard, we can see that this deeply ingrained and long perpetuated myth of black disingenuity has been a central element in attempts to justify slavery, as well as a whole array of racialized behaviors in the centuries after emancipation. But we are still left to wonder why scholars haven't stepped in with a more critically satisfying analysis of the relation between race and technology. The answer to that question lies at least partly in the evolution of the disciplines most concerned with those subjects.

In the United States, the history of technology emerged on a wave of post-World War II technological enthusiasm and economic ebullience. Perhaps naturally, it took on a celebratory character, emphasizing a triumphant technics, and Cold War politics reinforced that tendency. This kind of attention to great men and technological progress drove research into rather limited and exclusive channels that centered on big capital, complex technologies, and the small fragment of the population acting on that narrow stage. Inevitably, it dismissed all those who, to use Carroll Pursell's apt language, "were effectively barred by law, habit, and social expectation from the design and development stages of technical praxis."[5] It was a tale, in other words, of advantage and the successes that flowed from it.

This essentially conservative approach had its own theory. Brooke Hindle, one of the field's early spokesmen, claimed that there was a deep, interior logic to technology, crucial to the understanding of its meaning, and accessible only through rigorous study of its internal complexities.[6] That position argued the need for technical as well as historical training, and more selectively defined who and what was worth study. It took a new generation of historians to realize that technology is as much about process as about product, and that its history legitimately comprises the field as well as the factory, the home as well as the engineering site.

George Washington Williams published his *History of the Negro Race in America from 1619 to 1880* in 1882, though most people would date the origins of African-American history as a discipline to Carter Woodson's founding of the Association for the Study of Negro Life and History in 1915.[7] Still, it was not until the 1960s that African-American Studies became established in the academy, largely as a consequence of the civil rights movement and the research of a group of historians who wrote out of strong ideological conviction. The field that emerged continued a tradition of writing about race relations, implicitly if not explicitly, as a basis for political action. As it matured, however, scholars produced increasingly complex and subtle conceptual frameworks for analyzing race, including new understandings of agency—the ways in which men and women shape their own lives, even under disadvantageous circumstances.

These theoretical advances in both fields now open the way for an enriched history of technology and for new insights into the role of technology in African-American life. We have learned for a certainty that race is not a fixed, immutable concept—that definitions of who is white and who is black have changed with time, place, and circumstance. That technology is also a product of interest—political and ideological as well as economic—is also now widely accepted as an analytical point of departure. And we can begin to see that these subjects are more tightly connected than we imagined. Technology has long been an important element in the formation of racial identity in America. Whiteness and technological capability, Susan Smulyan points out, were usually seen as "natural" parts of each other, and as fundamental elements of masculinity.[8] By the end of the nineteenth century, these ideas had found widespread acceptance in such best-selling novels as *Trail of the Lonesome Pine* and *The Winning of Barbara Worth*, each subsequently made into a movie that featured a rugged, intelligent, problem-solving white engineer as the leading male character. An opposite calculus—the imputation of foolish incompetence in blacks, and thus the want of a key ingredient for independent manhood—found equally widespread acceptance. How and why these constructions were framed and how they interact thus becomes not just a good object of study, but a critical one.

There is a very reasonable argument to be made for the proposition that all discussions of race should go beyond the simple juxtaposition of black and white, and this is certainly true in the case of technology. But there is an equally persuasive logic for starting with African-Americans—because they are the classic American minority group, because they have been the focus of most American civil rights efforts, and because in their case American ideals of justice and equity are most specifically at issue.[9]

Yet, even with an enhanced appreciation of the complexities of these subjects and of their interrelatedness, we need also to be reminded that, although archival holdings and museum collections influence what historians study, people also make choices about what history gets written. Until feminist scholars created the analytical tools that revealed the women who had been there all along, historians could hardly imagine their existence.[10] Similarly, until very recently few historians have sought analytical tools that might link the study of African-Americans and technology. Just as it took new approaches to put women back into the story of America, so we now seek the means to write blacks back into the history of American technology.

To conceive such tools, we need to start not with African-Americans but with the ways in which white Americans have represented themselves. From the eighteenth century on, white Americans described themselves as an inventive people. They claimed to have a natural disposition for quick and novel solutions to the practical problems of life. That is what "Yankee ingenuity" meant—a self-attached label, applied early on.[11] And that distinct image, explicitly and repeatedly articulated over the next two centuries, was ideologically linked to the exploitation of the continent's natural resources as well as to the historic destiny white Americans imagined to be the just consequence of their political experiment. Democratic ideals would triumph by releasing the people's energies, and they would prosper by exploiting the resources that had been given them.

But that romantic vision was always framed in racial terms. European-Americans almost never considered the Africans among them, whether enslaved or enfranchised, to be capable of creative technical thought—and they translated that difference into an explicit point of contrast. Hundreds of examples illustrate that conviction, but they are all summed up in the sarcasm of a Massachusetts lawyer in a patent case when he said "I never knew a negro to invent anything but lies."[12] And even as colonial newspaper advertisements by the hundreds described the considerable craft skills of runaways, plantation owners insisted that enslaved Africans broke or misused their tools because they could not understand how to use them, not as deliberate acts of resistance.[13]

More than that, Ron Takaki points out, technology was perceived as the means by which people of color throughout the United States—African, Native American, Hispanic, and Asian—were to be subordinated to the grander purposes of American civilization.[14] All down these long decades, white, Anglo-Saxon, Protestant Americans made technology and the capacity for its skillful management central both to the task of nation-building and to the way they represented themselves. Just as plainly, they contrasted themselves to people of color, whom they judged incapable of such things. That's what Toni Morrison means by "Africanism," an explicit kind of marginalization against which privileged status can be defined.[15]

Our history with technology, then, has always been entangled in ideas about race. But the curious consequence is that we have written that history blind to color—as if accepting all those earlier assumptions about who was inventive and who was not, as if the ways in which a people thought about and used technologies were essentially irrelevant. This limited kind of understanding is currently under attack. The work of Takaki, Robert Rydell, and Michael Adas reveals the extent to which our historic concepts of technology and of our own technological prowess have been infused by racial ideology.[16] Even *Technology and Culture*—the principal journal on the history of technology—has started publishing articles that explicitly engage the issue of race and technology. One example, reprinted here, shows how rice cultivation in South Carolina and Georgia depended on knowledge brought to those places by enslaved Africans. We already knew from Peter Wood's work that lowland South Carolina planters preferred slaves from the rice-growing regions of Africa, and we knew that those slave owners were themselves originally ignorant of the techniques and processes of rice cultivation.[17] Now we can appreciate in more explicit detail the specifics of field layout, of irrigation methods, and of the technics of rice processing (all African imports), and what we learn directly challenges the notion that blacks contributed only their labor.[18] Another recent article in *Technology and Culture* describes the relation between race, changing technology, and work assignments at Bell Telephone, and shows how the technological displacement of labor was biased by color.[19] We always thought that happened; now we have a compelling analysis of the process. So, even if slowly, we begin to see that in our country technology and race have always been tied closely together, just as we begin to sense that those connections are much more intertwined and ubiquitous than we ever realized.

How can we throw even more light on these complexities? We might start by searching out all the black inventors who have never received appropriate credit. That approach not only gives the lie to the myth of disingenuity, but also offers the comfort of familiar ground. In this country we have always celebrated our inventors. We love telling success stories, imagining them to say something important about both our past and our future. And in fact we are now beginning to see some interesting work about black inventiveness. A good place to start is Portia James's *The Real McCoy*, an extensive catalog written to accompany an exhibit she developed at the Anacostia Museum of the Smithsonian Institution. In a revised form, her essay from that book is included in this volume.[20] Another source that will prove valuable is Rayvon Fouche's *Black Inventors in the Age of Segregation*, soon to be released by the John Hopkins University Press

Invention is, however, a problematic category of analysis. The patent system has always worked worst for the poor, who have had least access to its law and processes, and that proved doubly so for black inventors. Before 1865, they were even denied the right to a patent, so that slave owners could lay claim to the intellectual as well as the physical labor of their property. After 1865 blacks more often than not lacked the economic resources to develop their ideas into patentable or marketable form, and for that reason were often forced to sell their interest in inventions prematurely. The romance of invention focuses on the flash of creative insight, to use A. P. Usher's dramatic phrase, but financial rewards more often depend on the legal manipulation of patent rights—something else not easily managed from the margins of society.[21] Finally, it is important to realize that patents describe only a fragment of human inventive activity and are only a small part of the story of people's experiences with technologies. On the other hand, if that familiar model doesn't work very well, what new paradigms do we need in order to discover the connections we seek?

In fact, all it takes to reveal a much more richly populated and therefore more authentic history is to turn the older approach on its head. If, instead of concentrating on the production of new technology, we look equally hard at the worlds of labor and of consumption, then whole new casts of characters emerge. Let's start with work. After all, it was the work of African-Americans that created the rice, tobacco, and cotton economies of the South, and thus so much of America's eighteenth- and nineteenth-century agricultural wealth. Some of that labor also took place in factories, both before and after the Civil War. Charles B. Dew originally pointed out the crucial role played by skilled slave ironworkers in Richmond's Tredegar Ironworks, one of the South's largest industrial enterprises. In a subsequent analysis of smaller furnaces and forges in the great valley of Virginia, Dew revealed both the extent to which slave artisans (who couldn't go on strike) became the preferred work force and how their skills gave them some control over their own work assignments.[22] W. E. B. Du Bois, in *The Negro Artisan,* identified black workers with "considerable mechanical ingenuity" across a broad range of craft and manufacturing occupations.[23] World War I opened up new opportunities for black people in Northern factories, breaking the agricultural "job ceiling" (to use the words of Trotter and Lewis) and making blacks important contributors to the nation's industrial economy.[24]

Thinking about labor means establishing the historical worth of the work in which most people have always been engaged, and it means exploring more creatively the relations between work, technologies, and skill. I don't at all mean to suggest that we relegate the inventive imagination of Elijah McCoy or Granville Woods to a place of lesser historic importance. But if we intend a truly inclusive history, an argument Lonnie Bunch cogently advances in an essay reprinted here, then we have to take into account all those people whose most crucial encounter with machines and techno-logical systems takes place on the job. And surely it is the case that, in the normal, daily working of the world, skill and experience count for as much as abstract knowl-edge and formal training. What makes this fact important to us is that by defining technical knowledge and creativity in broad terms we immediately reveal hosts of African-Americans who had previously been excluded from the story. We find them planning the layout of South Carolina rice fields, creating pottery, fashioning the furniture now highly prized by collectors, using sewing machines, running and fixing cotton gins, molding iron in Henry Ford's assembly-line factories, and fishing in the ocean for schools of menhaden.[25]

Frederick Douglass understood the critical importance of these kinds of skills in American society, and more particularly he recognized the precise connection in our society between skill and manly status. In an 1848 letter to Harriet Beecher Stowe, he wrote: "We must become mechanics—we must build, as well as live in houses—we must make, as well as use furniture—we must construct bridges, as well as pass over them—before we can properly live, or be respected by our fellow men."[26]

Work has been an important theme of recent studies in African-American history. But in addition to the relation between labor and the creation of wealth, we also need to think about the connections between work and craft and about the affinity between craft skill and knowledge. Since the nineteenth century, engineering in this country has depended on a published literature and on advanced formal instruction that has included physics and mathematics. Craft skill depends on a different kind of knowl-edge, most of it unwritten and learned on the job. Apprenticeship—whether institu-tionalized or not—rests on emulation and repetitive practice in the interest of acquiring manual skills, and it is married to experience with the ways in which materials behave in different circumstances. Not only is this kind of knowledge complex and difficult to transfer; it gains importance when considered in the context of the history of American slavery, the formal acquisition of knowledge by slaves having been forbidden by law.

In the seventeenth century, there was little hesitancy at exploiting the technical talents of African labor. Edmund White, for instance, wrote in 1688 to Joseph Morton, twice governor of the Carolina colony: "let yr negroes be taught to be smiths, shoemakers & carpenters & bricklayers: they are capable of learning anything."[27] And learn they did. Robert Fogel estimates that by the eighteenth century 10 percent of all black women were engaged in cloth production, while upwards of half of all male slaves were employed in blacksmithing, leather-working, cooperage, and carpentry—all considered elite occupations, as were such subsequent pursuits as the management of steam engines, boilers, and other machinery.[28] Indeed, Fogel points out, plantations were industrial enterprises that employed advanced technologies and depended upon a wide variety of skills. A more complex division of labor yields more complex labor, and this fact is important as a corrective to the notion that enslaved blacks were ignorant of current technics and untouched by them.

Almost from the beginning, slavery in America was characterized by substantial technical talent and an elaborate occupational hierarchy. Moreover, planters encouraged the development of hierarchies, seeing it as a means of ensuring a tractable work force. As Fogel argues, "the critical decision made by the planters, the decision that allowed the eventual emergence of a many-sided and often quasi-autonomous slave society, was the switch from whites to slaves as the source of personnel for their various managerial and craft slots."[29] There were risks to this approach. Even as their owners encouraged legislation to prohibit the education of slaves, the teaching of craft skills often required some book learning. And knowledge combined with skill brought other contradictions. One planter ruefully observed that, analogous to the profit he made, these elite occupations rewarded their black practitioners with "an extra measure of pride."[30] So perhaps we shouldn't be surprised to learn that skilled craftsmen led most slave rebellions.

Identity through one's work has always been a fundamental part of our culture. Consider the maritime occupations, for example. Long before Frederick Douglass learned the ship caulker's trade, blacks—both free and unfree—worked at shipbuilding, as sailmakers, and as sailors.[31] On both sides of the Chesapeake, where waterways provided the dominant means of transportation, as well as the source of seafood and game, generations of African-American watermen and boat builders, down to the present, have practiced their crafts, as family histories are now beginning to reveal.[32]

Pursuing these kinds of investigations will amplify our understanding both of technology and of the diverse people engaged with it. And field is as relevant as factory; agriculture depends on a set of technologies, just as does fishing, mining, and forestry. Each also requires varied kinds of expertise in the management of its techniques, some of which, Barbara Garrity-Blake's essay provocatively suggests, can even be invisible in character.

Finally, examining the links between race and labor gives us more useful conceptual tools. Scholars have already noticed that while access to technology-related jobs has often been made a matter of color, that relationship has often changed as technologies have changed, and the assignments have also differed geographically. At one end of the range technologies displace labor, while at the other technologies create a demand for low-wage labor in high-risk conditions, some of which can include strikebreaking. Thus, new technologies constantly force the renegotiation of racialized work, and the whole history of that process remains to be written.[33]

An examination of the role of consumption similarly reveals a much more interesting picture of the relation between technology and race and promises an especially fruitful line of inquiry. Leaving aside the idea that consumers play a role in the design process, it can at least be said that outside of work, most of us encounter technologies as consumers—that is, through use. Patents, after all, have little historical importance if no one uses the thing invented, as happens more often than one might realize. Moreover, we know that people employ technologies differently. Black families in Atlanta used automobiles not only for work or personal convenience, but also to escape the humiliating experience of segregated systems of public transit—thus giving that technology a distinctly political purpose.[34] Indeed, Langdon Winner claims that technologies actually have politics embedded in their forms—an argument that might sound right to anyone familiar with the effects of technological unemployment on blacks.[35] And it works the other way, too. The furnaces and foundries at the Ford Motor Company, for example, replicated the social politics of the outside world when white workers decided that, regardless of pay scales, they would not work at such dirty jobs.[36]

Besides whatever practical ends or economic ambitions it serves, access to technology defines status and power. Electrical technicians of the late nineteenth century, in an attempt to establish their own primacy as experts in the fluid occupational demographics of that period, consistently belittled the technological competence of blacks and women.[37] People use technology that way—to maintain existing social arrangements, or to escape them. We can most clearly see how these behaviors and strategies play out in the case of novel technologies; Kathleen Franz's essay in this volume shows the rich research possibilities of this approach.

People also appropriate technologies for their own ends, which are often different than those originally intended. Women have been known to cook turkeys in dishwashers, using the drying cycle. A decade ago, young African-American musicians experimentally scratched a stylus across vinyl records to create an alternative sound that carried political and cultural meaning. Despite the subsequent commercialization of that sound, it is still a good example of people using their politics to rethink technologies.[38] And here we come back to that matter of representation. Bell hooks has focused our attention on "the politics of representation," and that issue bears with particular force for us here because it has been such a struggle for blacks to represent themselves as technically competent. Photography is an oblique but good example of the case. When black people used it, the camera became "a political instrument, a way to resist misrepresentation, as well as a means by which alternative images could be produced." The camera was crucial to the way they could picture themselves. It gave them a means to "participate fully in the production of images," regardless of class—an ability that was enormously important in a world where someone else usually controlled the ways in which African-Americans were represented. Photography became, as hooks puts it, "a powerful location for the construction of an oppositional black aesthetic."[39] This power to define reality provides a starting point from which to shape politics and culture differently. And it works two ways: cameras in black hands—just like the technology of music in black hands—allows for the creation of an alternative image, but that image also enables African-Americans to represent themselves as skillful in the management of those technologies.

Thus, the way we think about race is often shaped by the technology employed in the debate. That connection becomes clear if we look at communications media, and it tells us something important about the control of radio that "Amos 'n' Andy" was the first serial program broadcast nationally in the United States. Even though that particular show employed white actors who imitated Negro speech, in many other cases the networks depended on black artists for talent, an important reality for people of color. According to Stanley Crouch, African-Americans could "remember radio waves smacking down segregation and making the jazz and dance band broadcasts, for instance, national experiences in the most democratic sense possible."[40]

African-Americans have always been interested in new technologies. And, like most other Americans, they have believed in the regenerative powers of technology. Inevitably, they ascribed an array of possibilities to machines such as cars and airplanes—new economic opportunities, an escape from racism, the chance to claim a place for themselves in American society. But technologies that you cannot own are different. Blacks could and did buy phonograph records as a way of managing the content of that technology within their own homes. The content of radio, however, was much more difficult to control, as those "Amos 'n' Andy" broadcasts so blatantly revealed. Yet even in this case, the politics of radio technology allowed African-Americans at least one chance to manipulate programming for their own ends.

Barbara Savage tells the story in her recent book *Broadcasting Freedom*. The central character in this episode is Ambrose Calliver, Senior Specialist in Education of Negroes in the U.S. Department of Education.[41] Long interested in radio as a medium, Calliver wanted to develop a series of programs that would showcase African-American contributions to the nation's history, culture, and intellectual life. To that end, he adroitly linked technology and politics. First, he knew that the Roosevelt administration was concerned about the extent to which blacks would support the war effort, particularly since A. Philip Randolph—using the very rhetoric employed against Hitlerism to address the problems of racism at home—was threatening a march on Washington to protest discriminatory hiring in defense jobs. Calliver also appreciated the fact that government control of frequency allocation gave him leverage with network broadcasters, and he understood that this public character of radio made it especially suitable for educational content. Calliver skillfully manipulated these factors to push NBC into broadcasting a series called "Freedom's People," starting in the fall of 1941. Using an experienced science writer, he artfully orchestrated a message that began with comfortable, non-threatening music such as "Go Down Moses" and featured celebrated artists such as Paul Robeson. Then, in a conscious and deliberate way, Calliver progressed to shows on literature, science, discovery, invention, military service, and the skills of black workers—building his argument for the intellectual abilities, the inventive talents, the courage, and the capabilities of African-Americans, past and present.

Besides serving as a nice example of the intersection of race, politics, and technology, Calliver's radio series raises interesting questions that call for further study. We might ask, for instance, how race gets represented in communications media. African-Americans were anxious to counteract the vulgarity of "Amos 'n' Andy" and the way blacks were portrayed on programs like "The Jack Benny Show," but in casting "Freedom's People" Calliver and his advisors were also concerned not to have an announcer whose voice didn't sound black enough. So, one might ask, how do race and technology reconstruct each other in radio and in other media?

We can give meaning and form to our technologies as consumers, and we can shape their applications through politics, but it is important to understand that they do not come to us as a given. They are not the result of a neutral process, and they are certainly not the consequence of some inevitable technical logic. They are the result of choices, of social processes, and consequently they embody interests, positions, and attitudes. Steven Lubar puts it as follows: "Machines and technological systems, like other forms of material culture, render cultural and social relations visible, tangible, and artifactual, objectifying and externalizing them. Our machines reflect our culture and society."[42] More than that, even, one could argue that machines and technical processes—whether simple or complex—don't just mirror us, but rather they *are* our culture and society. In other words, all these objects, techniques, and systems, as well as the ways in which they are imagined, produced, employed, consumed, and experienced, are embodiments of the ways in which we think and act.

In their own work, historians of technology have demonstrated that technologies emerge from a rich mix of choices and constraints that are social, economic, political, and technical. But for all that effort, the notion of technology as a black box—something that comes to us in an inescapable form—is still widely popular. Consider, for example, a recent feature story in the *New York Times* about an array of small electronic devices, often installed and deployed without the knowledge of the car owner, that are increasingly being used to monitor people and their automobiles. In fact, these intrusive technologies are promoted by an array of interests that include insurance companies, fast food chains, and car rental agencies. Yet in speaking of their use, a faculty member at the University of Pennsylvania's law school concluded—as if the outcome were predetermined—that "technology goes forward and people are either forced to accept the loss of privacy or lose out on the benefits."[43] That casual observation, so reminiscent of the slogan of the 1933 Chicago World's Fair, "Science Finds, Industry Applies, Man Conforms," ignores both the contingent nature of technology and the unequal power relations in these transactions. And that is where including race in our analysis brings especially useful insights. Looking at technology from the vantage point of African-American history throws the issue of power into sharp relief. Technology may be socially constructed, but the players are not all on the same footing—a truth familiar to people of color, who have also long known that both its benefits and consequences are distributed unequally.

Once we understand technology in these broader terms, we can appreciate the fact that the history of technology in America must necessarily comprise a much larger segment of the population, black and white, than we have imagined. And this understanding of the material world we have created for ourselves, while more complex than our earlier ideas about these things, ultimately yields a truer, more empowering history. But this history will not write itself. The problem of sources is real; for want of written historical records, we know little of the enslaved African potter "Dave," of South Carolina, beyond the remarkable examples of his talent now housed in museums, and not much more of Thomas Day, the celebrated African-American furniture maker.[44] But, of course—it is worth repeating—what gets remembered is not simply a matter of documents but also of choice, of deciding what we will write about. And that decision often rests on what we imagine it possible to write about.   More and more, we are coming to see that there is an interesting and important history to be written about race and technology in America. Recent Ph.D. dissertations such as Linda Tucker's "Science at Hampton Normal and Agricultural Institute," Nina Lerman's on nineteenth-century industrial and vocational education in Philadelphia, Rayvon Fouche's on the African-American inventor Granville Woods, Jill Snider's "Flying to Freedom: African-American Visions of Aviation, 1910–1927," and Angela Lakwete's on the cotton gin are a few examples. But there is yet a great deal to be done. "Invisible Hands," an exhibit of black craftsmanship held at Macon, Georgia, suggested the possibilities of future work in material culture study. And anyone interested in pursuing the subject should begin with Theodore C. Landsmark's "Bibliography of African-American Material Culture," deposited at the Henry Francis du Pont Winterthur Library in Wilmington, Delaware.

Much is still to be discovered about the history of black scientific and technical educational institutions. Nina Lerman has written insightfully about race and education in nineteenth-century Philadelphia, and contributes an essay to this volume that suggests important larger themes on the subject, as well as an innovative conceptual framework. But while there were hundreds of technical colleges and institutes created to educate African-Americans, there is very little information about schools other than Tuskegee and Hampton. Amy Slaton's essay in this volume on more contemporary educational practice neatly outlines a research program that, besides providing an example of a successful grant proposal, might help us understand some of the roots of contrasting professional experiences between black and white engineers.

We also need a more complete exploration of African-American participation in the industrial exhibitions of the nineteenth and twentieth centuries—from regional fairs such as the Cotton States Expositions in Atlanta to national exhibitions such as the one held in Louisiana in 1904 and on to the great international expositions in Paris that Du Bois wrote about.[45] The Columbian Exposition in Chicago in 1893 presents especially rich materials for further examination. *The Reason Why the Colored American Is Not in the World's Columbian Exposition*, edited by Robert Rydell, is a good place to begin.[46]

At the local level, the study of African-American communities with technology in mind will reveal a wide range of technical knowledge and skills practiced by women and men, in their homes, stores, and shops. That was true of free black neighborhoods before 1865, and was certainly so in the urban centers of the later nineteenth and the twentieth century.[47]

We can see, then, that there is a great deal more to the interrelatedness of race and technology than scholars once thought, and a variety of interesting ways to come at this history. Upsetting as her story is, now that we know the dangers of overexposure to radiation, Rebecca Herzig's exploration of x-ray hair removal and skin whitening provides a provocative example of the varieties in analysis this subject offers. Furthermore, there is quite a substantial amount of rewarding material for study available both to teachers and students. The broad scope of Amy Bix's bibliographic essay reveals a surprising array of source materials and of research possibilities, and—together with the footnote references from the essays assembled here, many of which she incorporated into her essay—interested students will find all they need to make a start. Indeed, as we continue to explore the richness of this subject, the only surprise will be that we have waited so long to discover what lies at hand.

Notes

1. For the details of that correspondence, see Silvio Bedini, *The Life of Benjamin Banneker* (Scribner, 1971).

2. Frederick Law Olmsted, *A Journey in the Back Country, 1853–1854* (Schocken, 1970), reprint, p. 382.

3. The best source for the ways in which racial characteristics were defined in terms of scientific and technological accomplishment is Michael Adas, *Machines as the Measure of Men: Science, Technology, and Ideologies of Western Dominance* (Cornell University Press, 1989).

4. Robert Rydell, *All the World's a Fair: Visions of Empire at American International Expositions, 1876–1916* (University of Chicago Press, 1984).

5. Carroll W. Pursell, Listening for the Silences, position paper presented at a workshop on Technology and the African-American Experience, Atlanta, February 4, 1994.

6. Brooke Hindle, *Technology in Early America: Needs and Opportunities for Study* (University of North Carolina Press, 1966).

7. See Robert L. Harris, "The Flowering of Afro-American History," *American Historical Review* 92 (1987), December: 1150–1161.

8. Susan Smulyan, The Social Construction of Race in the United States, a position paper presented at workshop on Technology and the African-American Experience, Atlanta, February 4, 1994.

9. This case is well made by Ron Takaki on p. 7 of *A Different Mirror: A History of Multicultural America* (Little, Brown, 1993).

10. For examples of how feminist scholars have changed the history of technology, see Judy Wajcman, *Feminism Confronts Technology* (Pennsylvania State University Press, 1991); Angela

N. H. Creager et al., *Feminism in Twentieth-Century Science, Technology, and Medicine* (University of Chicago Press, 2001); *Technology and Culture* 38 (1997), special edition edited by Nina Lerman, Arwen Palmer Mohun, and Ruth Oldenziel.

11. A good example of this notion of a predisposition toward inventiveness, as well as a telling case of postwar technological enthusiasm, can be found in John A. Kouwenhoven, *Made in America: The Arts in Modern American Civilization* (Branford, 1948).

12. W. E. Burghardt Du Bois, "The American Negro at Paris," *American Monthly Review of Reviews* 22 (1900), p. 576.

13. Ira Berlin, *Many Thousands Gone: The First Two Centuries of Slavery in North America* (Harvard University Press, 1998), p. 120.

14. Ron Takaki, *Iron Cages: Race and Culture in 19th-Century America* (Oxford University Press, 1990).

15. Toni Morrison, *Playing in the Dark: Whiteness and the Literary Imagination* (Harvard University Press, 1992).

16. Adas, *Machines as the Measure of Men*; Robert Rydell, *All the World's a Fair* (University of Chicago Press, 1987).

17. "Literally hundreds of black immigrants were more familiar with the planting, hoeing, processing, and cooking of rice than were the European settlers who purchased them." (Peter H. Wood, *Black Majority: Negroes in Colonial South Carolina, From 1670 through the Stono Rebellion*, Norton, 1974, p. 61) See also Peter H. Wood, "'It Was a Negro Taught Them': A New Look at African Labor in Early South Carolina," *Journal of Asian and African Studies* 9 (1974): 160–179.

18. Judith Carney, "Landscapes of Technology Transfer: Rice Cultivation and African Continuities," *Technology and Culture* 37 (1996), January: 5–35.

19. Venus Green, "Goodbye Central: Automation and the Decline of 'Personal Service' in the Bell System," *Technology and Culture* 36 (1995), October: 912–949.

20. Pursell, Listening for the Silences.

21. Abbott Payson Usher, *A History of Mechanical Inventions* (McGraw-Hill, 1929). On the manipulation of patents, see also Carolyn C. Cooper, *Shaping Invention: Thomas Blanchard's Machinery and Patent Management in Nineteenth-Century America* (Columbia University Press, 1991).

22. Charles B. Dew, *Ironmaker to the Confederacy: Joseph Anderson and the Tredegar Iron Works* (Yale University Press, 1966), pp. 29–31; Dew, *Bond of Iron: Master and Slave at Buffalo Forge* (Norton, 1994), pp. 67–70.

23. W. E. Burghardt Du Bois, *The Negro Artisan* (Atlanta University Press, 1902), p. 188.

24. Joe W. Trotter and Earl Lewis, eds., *African Americans in the Industrial Age: A Documentary History, 1915–1945* (Northeastern University Press, 1996), p. 1.

25. John Michael Vlach has provided the best information on African-American craft workers. See, for example, his book *The Afro-American Tradition in Decorative Arts* (Cleveland Museum of Art Press, 1978). Barbara Garrity-Blake's *The Fish Factory: Work and Meaning for Black and White Fishermen of the American Menhaden Industry* (University of Tennessee Press, 1994) is a fascinating study of the intersection of work, mechanism, and social relations.

26. "Proceedings of the 1853 Colored National Convention at Rochester, New York" (Frederick Douglass to Harriet Beecher Stowe, March 8, 1848), in *Minutes of the Proceedings of the National Negro Conventions, 1830–1864*, ed. H. Bell (Arno, 1969). Another useful source of information on the subject of work and skill is *A Guide to the Microfilm Edition of Slavery in Ante-Bellum Southern Industries*, ed. M. Shipper (University Publications of America, 1997).

27. Peter H. Wood, *Black Majority: Negroes in Colonial South Carolina From 1670 through the Stono Rebellion* (Norton, 1974), pp. 43–44.

28. Robert William Fogel, *Without Consent or Contract: The Rise and Fall of American Slavery* (Norton, 1989), p. 50.

29. Ibid., p. 58.

30. Berlin, *Many Thousands Gone*, p. 137.

31. Jeffrey Bolster, *Black Jacks: African American Seamen in the Age of Sail* (Harvard University Press, 1997).

32. Harold Anderson, "Black Men, Blue Waters: African Americans on the Chesapeake," *Maryland Marine Notes* 16 (1998), March-April: 1–3, 6–7.

33. One example of this literature is Jaqueline Jones, *American Work: Four Centuries of Black and White Labor* (Norton, 1998). For an account of the ways in which technologies create demand for low wage, high-risk jobs see Armando Solorzano and Jorge Iber, "Digging the 'Richest Hole on Earth': The Hispanic Miners of Utah, 1912–1945," *Perspectives in Mexican American Studies* 7 (2000): 1–27.

34. Blaine A. Brownell, "A Symbol of Modernity: Attitudes toward the Automobile in Southern Cities in the 1920s," *American Quarterly* 24 (1972), March, p. 35.

35. Langdon Winner, *The Whale and the Reactor: A Search for Limits in an Age of High Technology* (University of Chicago Press, 1986).

36. Joyce S. Peterson, "Black Automotive Workers in Detroit, 1910–1930," *Journal of Negro History* 64 (1978): 177–190. See also Warren Whatley, African Americans, Technology, Work and the Reproduction of Racial Differencing, unpublished research paper, 1994.

37. Carolyn Marvin, *When Old Technologies Were New: Thinking About Electrical Communication in the Late Nineteenth Century* (Oxford University Press, 1988).

38. Tricia Rose, *Black Noise: Rap Music and Black Culture in Contemporary America* (Wesleyan University Press, 1994).

39. Bell hooks, "In Our Glory," in *Picturing Us*, ed. D. Willis (New Press, 1994), p. 49.

40. Stanley Crouch, *The All American Skin Game, or The Decoy of Race* (Vintage, 1995), p. 110.

41. Barbara Dianne Savage, *Broadcasting Freedom: Radio, War, and the Politics of Race, 1938–1948* (University of North Carolina Press, 1999).

42. Steven Lubar, Technology and Race, position paper presented at workshop on Technology and the African-American Experience, Atlanta, February 5, 1994.

43. *New York Times*, October 25, 2001.

44. See Vlach, *The Afro-American Tradition in Decorative Arts*.

45. See Du Bois, "The American Negro at Paris," and Philip S. Foner, "Black Participation in the Centennial of 1876," *Phylon* 39 (1978): 283–295. Nicholas Murray Butler of Columbia

University also presented an exhibit of American higher education at the Paris exposition, irresistibly suggesting a comparison with Du Bois's experience there.

46. Robert Rydell, ed., *The Reason Why the Colored American Is Not in the World's Columbian Exposition* (University of Illinois Press, 1999).

47. For an example of a study of free black communities, see James O. Horton, *Free People of Color: Inside the African American Community* (Smithsonian Institution Press, 1993). See also his more recent studies *In Hope of Liberty* (Oxford University Press, 1997) and *Black Bostonians* (Holmes & Meier, 1999).

Landscapes of Technology Transfer: Rice Cultivation and African Continuities

Judith Carney

By the mid 1700s a distinct cultivation system, based on rice, rimmed the Atlantic basin. The eastern locus of rice cultivation extended inland from West Africa's upper Guinea coast. To the west the system flourished in the southeastern United States, principally along the coastal plain of South Carolina and Georgia (figure 1). On both sides of the Atlantic, rice growing depended on African labor. West African farmers planted rice as a subsistence crop on small holdings, with surpluses occasionally marketed, while the southeastern United States depended on a plantation system and West African slaves to produce a crop destined for international markets.

While rice cultivation continues in West Africa today, its demise in South Carolina and Georgia swiftly followed the abolition of slavery. The year 1860 marked the apogee of the antebellum rice economy. Total U.S. production reached 187.2 million pounds, with South Carolina accounting for 63.6 percent of the total and Georgia an additional 28 percent.[1] Abolition doomed this rice plantation system by liberating some 125,000 slaves who grew rice along nearly 100,000 acres of coastal plain, the property of about 550 planters.[2]

The South's most lucrative plantation economy continued to inspire nostalgia well into the twentieth century, when the crop and the princely fortunes it delivered remained no more than a vestige of the coastal landscape. Numerous commentaries documented the life ways of the planters, their achievements, as well as their ingenuity in shaping a profitable landscape from malarial swamps.[3] These accounts never presented African slaves as having contributed anything but their unskilled labor. The 1970s witnessed a critical shift in perspective, as historians Converse Clowse and Peter Wood drew attention to the skills of slaves in the evolution of the South Carolina economy. Clowse, writing in 1971, revealed the importance of skilled African labor in ranching and forest extractive activities during the early colonial period. Wood's careful examination of the role of slaves in the Carolina rice plantation system during the same period, published in 1974, showed that slaves contributed agronomic expertise

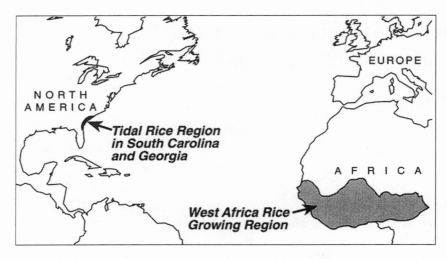

Figure 1
Rice cultivation along the Atlantic basin, 1760–1860.

as well as skilled labor to the emergent plantation economy. Wood's argument rested on the presence of slaves in South Carolina from the onset of settlement in 1670, early accounts suggesting that slaves produced their own subsistence crops, and the contrast between a lack of prior rice farming knowledge among the English and the French Huguenot planters and the knowledge and skill of their African slave work force.[4] Daniel Littlefield later built on Wood's thesis by discussing the antiquity of African rice farming practices and by demonstrating that more than 40 percent of South Carolina's slaves during the colonial period originated in West Africa's rice cultivation zone.[5]

While this scholarship has resulted in a revised view of the rice plantation economy as one of both European and African culture, the agency of African slaves in its evolution is still debated. Current formulations question whether planters recruited slaves from West Africa's rice coast to help them develop a crop whose potential they discovered independently or whether African-born slaves initiated rice planting in South Carolina by teaching planters to grow a preferred food crop. The absence of archival materials that might document a tutorial role for African slaves is not surprising given the paucity of records available in general for the early colonial period, and because racism over time institutionalized white denial of the intellectual capacity of bondsmen. An understanding of the potential role of slaves requires other forms of historical inquiry.

This essay combines geographical and historical perspectives to examine the likely contributions of African-born slaves to the colonial rice economy. A spatial approach is used to focus attention on the principal micro-environments planted to rice on both sides of the Atlantic, as well as on the techniques developed for soil and water management during the colonial period. The period 1670–1770 is crucial for examining these issues since it spans the initial settlement of South Carolina by planters and slaves as well as the expansion of tidal (tidewater) rice cultivation into Georgia. By analyzing the spatial and agronomic (i.e., land management) parameters of rice cropping systems, this cross-cultural analysis emphasizes linkages between culture, technology, and the environment that traversed the Middle Passage with slaves.

The discussion has four parts. The first section examines the geographical and historical context for rice cultivation in the Atlantic basin, while the water and soil management principles underlying the three major West African rice systems are discussed in greater detail in the second part. The discussion shifts in the third section to South Carolina and Georgia, where the rice economy unfolded overtime from rain fed to inland swamp production and culminated in the tidal system. The concluding section raises several questions about the issue of technology development and transfer while suggesting a lingering Eurocentric bias in historical reconstructions of the agricultural development of the Americas.

The Geographical and Historical Context of Rice Cultivation along the Atlantic Basin

Some 3,500 years ago, West Africans domesticated rice (*Oryza glaberrima*), independently of the Asian cultivar (*Oryza sativa*), along the flood plain and inland delta of the upper and middle Niger River in Mali.[6] The rice-growing area of West Africa, depicted in figure 2, extends along the coast from Senegal to the Ivory Coast and into the interior through the savanna along river banks, inland swamps, and lake margins. Within this climatically and geographically diverse setting, two secondary centers of rice domestication emerged: a flood-plain system located along river tributaries north and south of the Gambia River (bounded on the north by the river Sine and on the south by the river Casamance, both in Senegal); and another, rain-fed system found farther south in the forested Guinea highlands where rainfall reaches 2,000 mm per year. By the end of the seventeenth century rice had crossed the Atlantic basin to the United States, appearing first as a rain-fed crop in South Carolina before its diffusion along river flood plains and into Georgia from the 1750s.

Many similarities characterized rice production on both sides of the Atlantic basin. The most productive system in South Carolina developed along flood plains, as in West Africa. Precipitation in each region follows a marked seasonal pattern, with rains generally occurring during the months from May to October. Rice cultivation flourished

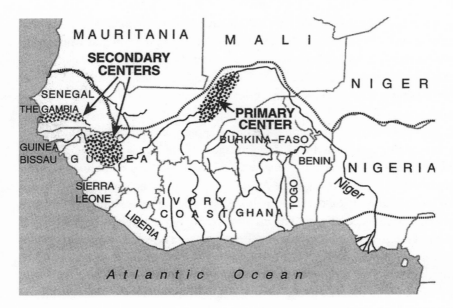

Figure 2
Centers of origin of African rice *(Oryzo glabemma)*.

in South Carolina and Georgia under annual precipitation averages of 1,100–1,400 mm, figures that represent the midrange of a more varied rainfall pattern influencing West African rice cultivation.[7] In the West African production zone, precipitation increases dramatically over short distances in a north-to-south direction, with much less variation occurring over greater distances from east to west. Thus, in the Gambian secondary center of domestication, semiarid (900 mm annual precipitation) conditions prevail, as in the Malian primary center, while farther south in Guinea-Bissau and Sierra Leone precipitation exceeds 1,500 mm per year.[8]

The topography of the rice-growing region on both sides of the Atlantic presents a similar visual field. The coastline is irregularly shaped and formed from alluvial deposits that also create estuarine islands. Rivers carry freshwater downstream on their way to the sea, resulting in tidal flows over the flood plain. The steeper descent from the piedmont in South Carolina and Georgia delivers freshwater tidal flows to flood plains just 10 miles from the Atlantic coast. But the less-pronounced gradient of West Africa's rice rivers means that freshwater tides meet marine water much farther upstream from the coast. Saltwater constantly menaces the downstream reaches of rivers like the Gambia, where salinity permanently affects the lower 80 km but seasonally intrudes more than 200 km upstream.[9] The advance of saltwater in coastal estuaries, however, failed to discourage West Africans from adapting rice cultivation to this environment. Farther south of the Gambia River, where rainfall exceeds 1,500 mm, rice planting occurs in an even more challenging environment—on tidal flood plains formed under marine water influence. The underlying potential acid-sulfate soils depend on water saturation to prevent them from oxidizing and developing the acidic condition that would preclude continued planting.[10] West African rice farmers avoid this problem by constructing an elaborate network of embankments, dikes, canals, and sluice gates to bar marine water while capturing rainfall for cultivation.

Rice cultivation continues throughout West Africa today under conditions similar to those that prevailed at the onset of the Atlantic slave trade. When Islamic scholars followed preexisting overland trade routes to the Malian empire in the fourteenth century, they arrived in the heart of West African rice domestication, where food surpluses had sustained empire formation from the ninth century.[11] Their earliest commentaries mention the crop's widespread cultivation.[12] Description of West African rice systems came later, with the arrival of European vessels along the Atlantic coast from 1453. Dependence on marine navigation brought the Portuguese into early contact with the rice cultivation systems developed along coastal estuaries (mangrove rice) and tidal rivers.[13]

As the Atlantic slave trade gained momentum in the region over the next century, rice cultivation on the littoral drew repeated interest and comment. In 1594, long before the permanent settlement of South Carolina, André Alvares de Almada provided the first detailed description of the mangrove rice system that continues to characterize coastal estuary production south of the Gambia River. He noted the use of dikes to impound rainwater for seedling submergence and desalination, ridging to improve soil aeration, and transplanting.[14] De Almada's description leaves no doubt as to the existence of sophisticated water and soil management techniques from the earliest period of contact with Europeans. One eighteenth-century slaver captain, Sam Gamble, found this system so elaborate that he provided a diagram of the field layout in conjunction with his description of water-management techniques (figure 3).[15]

Discussion of the upland and inland swamp cultivation systems away from coastal and riverine access routes first appeared around 1640 in a manuscript published by an Amsterdam geographer, Olfert Dapper. Relying on information supplied by Dutch merchants operating during the early seventeenth century in the region currently known as Sierra Leone and Liberia, Dapper described a rice farming system in which cultivation occurred in distinct micro-environments along a lowland-to-upland landscape gradient that included inland swamps as well as uplands: "Those who are hardworking can cultivate three rice-fields in one summer: they sow the first rice on low ground, the second a little higher and the third . . . on the high ground, each a month

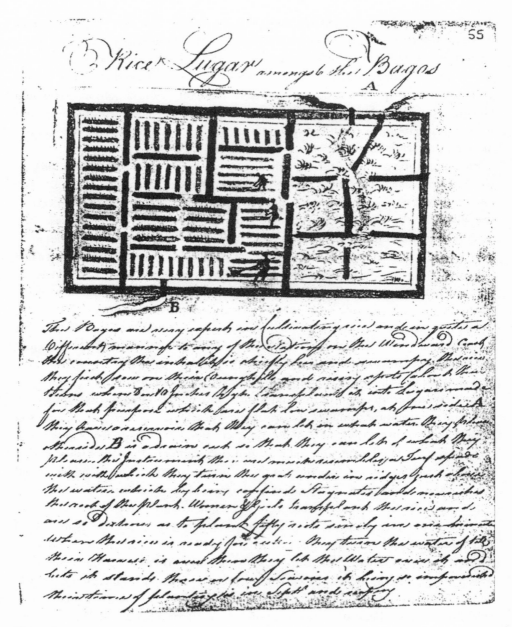

Figure 3
Sam Gamble's eighteenth-century diagram of a coastal West African rice field. Source: D. Littlefield, *Rice and Slaves* (Baton Rouge, 1981). Reprinted with permission.)

after the previous one, in order not to have all the rice ripe at the same time. This is the commonest [sic] practice throughout the country. . . . The first or early rice, sown in low and damp areas . . . the second, sown on somewhat higher ground . . . the third, sown on the high ground."[16]

Additional observations of these systems came later, during the mid-eighteenth century, when Europeans financed overland expeditions for exploration, trade, and science.[17] The growing dispersal of Europeans into the West African interior during the nineteenth century brought more detailed commentaries on forest clearance for planting rain-fed rice and the field's subsequent rotation for cattle grazing, as well as the use of earthen reservoirs in inland swamps for water impoundment against drought.[18]

The Agronomic and Technological Bases of West African Rice Systems

Even though higher-yielding Asian rice (*O. sativa*) varieties and pump-irrigation systems now dominate throughout the West African rice zone, the production systems that predate the Atlantic slave trade persevere in the region today. Several researchers have favorably compared the diversity of the West African rice systems to those in Asia, especially noting the African crop's production under quite different rainfall patterns, soils, farming systems, and land types.[19] A recent study underscored the diversity of micro-environments and land-management practices characteristic of the African systems by identifying eighteen different environments planted in the West African rice zone.[20] This article emphasizes the main features of African production systems that potentially bear on the evolution of the rice plantation economy in South Carolina and Georgia.[21] Prioritized for discussion are the type of water regime(s) utilized for production, the underlying agronomic techniques, and each system's relationship to yield and labor availability.

Three principal water regimes influence West African rice planting: rainfall, groundwater, and tides. The resultant rice systems are respectively known as upland, inland swamp, and tidal production.[22] The source of water for rice planting also reflects the cropping system's position along a lowland-to-upland landscape gradient. West African rice systems occur along a continuum of changing ecological conditions where one or more moisture regimes are manipulated for production.[23] Figure 4, a transect of rice cultivation along the central Gambia River, illustrates this landscape gradient.

The long-standing practice of planting rice sequentially along a landscape continuum from low to high ground confers several advantages, as Dapper noted in the early seventeenth century.[24] By manipulating several water regimes, farmers initiate and extend rice growing beyond the confines of a precipitation cycle. In so doing they even

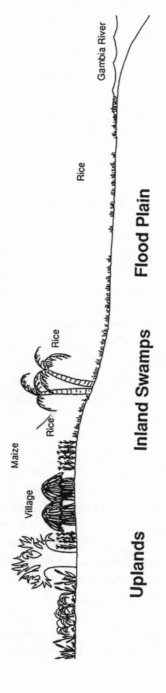

Figure 4
Lowland-to-upland landscape gradient of rice cultivation along the Gambia River.

out potential labor bottlenecks, since cropping demands (sowing, transplanting, weeding, and harvesting) are staggered through different periods of the agricultural season. By relying on several sources of water and several cropping environments, farmers enhance subsistence security by minimizing the risk of the rice crop completely failing in any given year.

Of the three forms of rice cultivation, the upland system depends strictly on precipitation; for this reason it is usually planted last in a rice farming system. The designation "upland rice" derives from planting the crop at the top of the landscape gradient, which may rest a mere 100 feet above sea level. Where rainfall reaches at least 1,000 mm, West African farmers commonly plant the crop in rotation with cattle grazing. Once cleared of vegetation, and with surface debris burned off, fields are planted to upland rice. Following harvest, the rice field converts to cattle grazing. The animals feed on the crop residues, their manure fertilizing the soil. The variability of rainfall patterns both within and between agricultural seasons results in generally lower yields (often below 1 ton per hectare) with upland rice cultivation than in the other two systems. Despite this risk and the initial effort expended for land clearance, upland rice demands less labor than the other West African production systems.

The second system, cultivation in inland swamps, overcomes the precipitation constraints of upland production by capturing groundwater from artesian springs, perched water tables, or catchment run-off. "Inland swamps" actually refers to a diverse array of micro-environments which include valley bottoms, low-lying depressions, and areas of moisture-holding clay soils. The broad range of inland swamps sown to rice reflects farmers' sophisticated knowledge of soils, particularly moisture-retention properties, and their effective methods to impound water for supplemental irrigation. Where high ground water tables keep inland swamps saturated, rice planting may begin before the rains, continue beyond the wet season, and take place under low rainfall conditions (below 900 mm per year).

But planting rice in inland swamps requires careful observance of topography and water flow. Farmers often construct bunds, small earthen embankments, around the plot to form a reservoir for capturing rainfall or stream runoff. The practice keeps soils saturated through short-duration dry cycles of the cropping season. This form of supplemental irrigation drew the interest of the French explorer René Caillié, who in 1830 noted: "As the country is flat, they take care to form channels to drain off the water. When the inundation is very great, they take advantage of it and fill their little reservoirs, that they may provide against the drought and supply the rice with the moisture it requires."[25] If excess flooding threatens the rice crop, the plot can be quickly drained by puncturing the bund.

Farmers sometimes improve drainage and aeration in inland swamp plots by ridging the soil. Rice seedlings are then either sown directly atop the ridges or transplanted, the latter method being favored when water-logging poses a risk to seedling development. In such cases, the young plant is established near the village and transplanted to the inland swamp about three weeks later. Though more labor-intensive, transplanting increases the plot's productivity. The process of pulling up the seedlings strengthens the root system and promotes tillering, which can increase rice yields by as much as 40 percent over direct-seeded plots.[26] Such practices and the use of supplemental groundwater resources combine in inland swamp rice production to improve subsistence security and increase rice yields, which generally exceed 1 ton per hectare.

The remaining African production system, tidal rice, occurs on flood plains of rivers and estuaries. Dependent on tides to flood and/or drain the fields, cultivation involves a range of techniques from those requiring little or no environmental manipulation (planting on freshwater flood plains) to ones demanding considerable landscape modification (cultivation along coastal estuaries, known as mangrove rice). Tidal rice cultivation embodies complex hydrological and land management principles that prove especially pertinent for examining the issue of African agency in the transfer of rice cultivation to the Americas.

Tidal rice cultivation occurs in three distinct flood-plain environments: along freshwater rivers, seasonally saline rivers, and coastal estuaries with permanent marine water influence. The first two involve similar methods of production—letting river tides irrigate the rice fields—while the third system combines principles of each major rice system for planting under problematic soil and water conditions.

The riverine flood plain in the first two systems actually includes two microenvironments: the area alongside a river irrigated by diurnal tides, and its inner margin, reached only during full moon tides, where the landscape gradient begins to rise (see the landscape continuum illustrated in figure 4). The inner flood plain's position along this gradient means that the crop relies in part on rainfall for water requirements. As the inner flood plain receives only occasional tidal flooding, rice varieties are frequently directly sown. But the flood plain flooded daily requires transplanting so that seedlings first reach sufficient height to withstand tidal surges. Both flood plain crops mature from moisture reserves captured in the alluvium during flood recession.

Similar topographic distinctions and agronomic techniques apply in the second type of riverine flood plain cultivation, which involves careful observation of salt and freshwater dynamics in order to plant areas under the seasonal influence of marine water. As rains discharge freshwater into West African rivers after the onset of the wet season, the saltwater interface retreats downstream. Rice cultivation takes place along riverine stretches that experience at least three months of freshwater (the maturation time of the fastest-growing seed varieties). This second type of tidal system requires less labor than the freshwater one, since seasonal saltwater conditions depress weed growth, but yields are similar, averaging between 1 and 2 tons per hectare, with the lower range found on the inner flood plains.[27]

Mangrove rice, the third form of tidal cultivation, takes place along coastal estuaries and represents the most sophisticated West African production environment. The principles underlying this system have not been sufficiently conceptualized by historians of rice development in South Carolina who have looked to West Africa for potential influences.[28] Their comparisons of rice production on both sides of the Atlantic basin understandably focus on the tidal freshwater system that sustained the lucrative antebellum economy of South Carolina and Georgia. But an emphasis on one production environment misses Dapper's seventeenth-century insight that rice planting occurs in distinct production environments along a landscape gradient. By separating out for analysis just one among the multiple environments that typically characterize a rice farming system, scholars only glimpse a fraction of the agronomic techniques and specialized knowledge that inform West African cultivation.

Thus, in emphasizing freshwater flood plain production, Littlefield correctly concludes that the African system involves minimal landscape manipulation.[29] However, rice production in tidal estuaries, on the other hand, demands considerable landscape modification and, sometimes, inter-village cooperation to manage the extensive water control system.[30] One important outcome of this emphasis on the West African tidal river system in cross-cultural comparison is to leave unquestioned the assumption that Europeans provided the technological basis to the South Carolina tidewater system.[31] The case for African agency in introducing the sophisticated soil and water management infrastructure to South Carolina flood plains dramatically improves by detailing the mangrove rice system.

West African rice production in tidal estuaries occurs south of the Gambia River in areas of permanently saline water conditions where annual rainfall generally exceeds 1,500 mm. These are environments mantled by extensive stretches of mangroves whose aerial roots trap fertile alluvium swept over the littoral by marine tides. The organic matter deposited on these soils makes them among the most fertile of the West African rice zone, but they require considerable care to prevent oxidation and their transformation into acid-sulfate soils. By manipulating several water regimes and developing the infrastructure for its control, a highly productive rice system results.

Preparation of a tidal rice field begins with site selection and the construction of an earthen embankment parallel with the coast or riverine arm of the sea (figure 5).

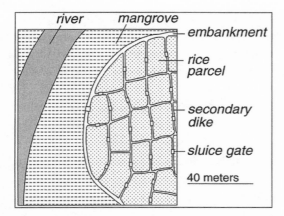

Figure 5
Tidal rice-production system, Casamance, Senegal. Adapted from P. Pélissier, *Les paysans du Senegal* (Saint-Yrieix, 1966).

Frequently more than a meter's height and width (the dimensions needed to block the entry of marine tides onto the rice field), the embankment stretches for several kilometers, sometimes threading together rice fields of different villages. A stand of mangroves often is left in place between the estuary and the embankment to reduce tidal force. The void left by soil removal for the embankment establishes the location of the principal drainage canal. A series of secondary embankments (dikes) are then formed perpendicular to the main one in order to divide the perimeter into the individual rice fields. The mangrove rice system achieves a dual purpose with water control. It captures rainfall for irrigation while storing water for the controlled floodings that drown unwanted weeds. Sluices built into the dikes facilitate water control through the canals for either irrigation or drainage. Fitted with valves made from hollow tree trunks and plugged with palm thatch, the dike sluices drain into a more substantial one located in the principal embankment. The principal sluice is sometimes fashioned from an old canoe with a board vertically positioned like a rudder to control water flow.[32]

Once enclosed, the field is flooded by impounding rainwater and evacuating it from the field at low tide. Rainfall (and sometimes seasonal freshwater springs) leaches out the salts, which low tides help evacuate into the estuary.[33] It takes years to desalinate a mangrove field before cultivation can ensue. Each season, as farmers await the rains that will rinse away residual dry season salts, they establish rice in nurseries near the village where the plants can be hand watered. After about a month's growth the seedlings are transplanted atop the rice field's ridges, a practice that promotes protection from residual salinity. At this point, the mangrove rice field reverts to rain-fed cultivation. Harvest occurs about four months later, the crop ripening from accumulated moisture reserves after the rains cease. Farmers annually renew soil fertility during the dry season by periodically opening the sluices to tidal marine water. This action prevents the soil acidification that leads to acid-sulfate formation and permits deposit of organic matter. In the month or so preceding cultivation, the sluices once again remain closed to block the entry of saltwater. Farmers prepare the plot for the new cultivation cycle by layering the ridges with accumulated deposits of swamp mud.

The creation of a mangrove rice system demands considerable labor for perimeter construction and desalination over a period of several years, as well as much effort annually to maintain the system's earthworks. But the reward for such monumental labor is yields that frequently exceed 2 tons per hectare. Besides displaying the range of soil and water management techniques developed for rice cultivation, the mangrove system illustrates a preexisting West African familiarity with the sophisticated earthworks infrastructure long associated with the South Carolina and Georgia tidal rice plantation.

The complex soil and water management principles embodied in planting rice along a landscape gradient in interconnected environments illustrate the ingenuity that characterized West African rice production. Numerous affinities exist with the rice systems of South Carolina and the process of technology development in tidewater rice, the antebellum era's quintessential production system.

The Temporal and Spatial Discontinuities of Rice Cultivation in South Carolina and Georgia

By 1860 rice cultivation extended over 100,000 acres along the eastern seaboard from North Carolina's Cape Fear River to the St. Johns River in Florida and inland for some 35 miles along tidal waterways (figure 6).[34] The antecedents of the rice plantation economy date to the first century of South Carolina's settlement (1670–1770), and especially to the decades prior to the 1739 Stono slave rebellion. Rice cultivation systems analogous to those in West Africa, using identical principles and devices for water control, were already evident in this period. Dramatic increases in slave imports during the eighteenth century facilitated the evolution and commoditization of the South Carolina rice economy. The process of technology development, moreover, occurred in tandem with the emergence of the task labor system that distinguished coastal rice cultivation. Colonial rice production shifted respectively from reliance on rainfall to inland swamps and, from mid-century, to the tidal (tidewater) cultivation system that characterized the antebellum era.

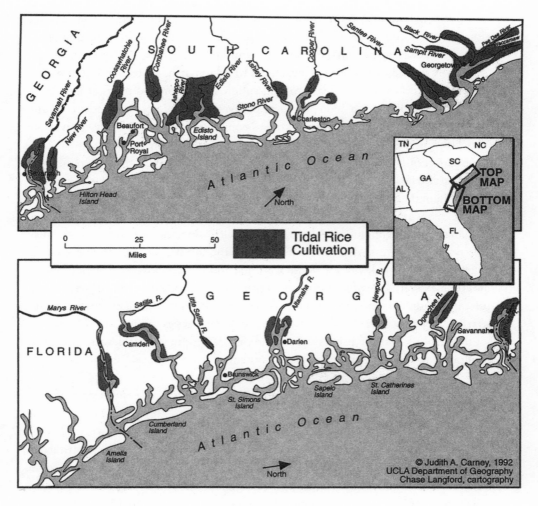

Figure 6
Tidewater rice cultivation in South Carolina and Georgia.

This section presents an overview of the material conditions and historical circumstances within which rice production developed in South Carolina. The technical changes marking the evolution of the colonial rice economy illuminate three issues that bear on comparative studies of technology and culture: first, the need in cross-cultural analysis to examine the technical components of production as part of integrated systems of knowledge and not merely as isolated elements; second, the extent to which superior social status coincides with a superiority in knowledge; and third, the relationship between technical expertise, patterns of labor utilization, and technological change.

Slaves accompanied the first settlers to South Carolina in 1670. Within two years they formed one-fourth of the colony's population, their numbers surpassing whites as early as 1708.[35] Rice cultivation appears early in the colonial period, with planting occurring in numerous environments. The earliest reference to the cereal's cultivation dates to 1690, when plantation manager John Stewart claimed to have successfully sown rice in twenty-two different locations.[36] Just five years later cultivation efforts culminated in South Carolina's first recorded rice exports: one and one-fourth barrels shipped to Jamaica.[37] In 1699 exports reached 330 tons, and during the 1720s rice emerged as the colony's leading trade item.[38] Years later, in 1748, Governor James Glen drew attention to the significance of rice experimentation during the 1690s for the subsequent unfolding of the South Carolina economy.[39]

The growing emphasis on rice exports from the turn of the century resulted by the 1740s in the documented presence of all three principal West African production systems: upland, inland swamp, and tidal.[40] Each dominated a specific phase in the South Carolina rice economy. As rice was gradually commodified during the eighteenth century, the numerous production environments mentioned by Stewart no longer characterized the pattern of rice cultivation. Instead, planting occurred in specific environments selected for emphasis at different moments in the crop's evolution as a commodity.

Upland rice production received initial emphasis in the eighteenth century for its complementarity with the colony's early economic emphasis on stock raising and forest product extraction. Slave labor underpinned this agropastoral system, which involved clearing forests, the production of naval stores (pine pitch, tar, and resin), cattle herding, and subsistence farming.[41] The export of salted beef, deerskins, and naval stores generated the capital for additional slaves. With the dramatic increase in slave imports (from 3,000 in 1703 to nearly 12,000 by 1720) rice cultivation shifted to the inland swamp system.[42]

The higher-yielding inland swamp system represented the first attempt at water control in South Carolina's rice fields but demanded considerable labor for clearing the cypress and gum trees and developing the network of bunds and sluices necessary for converting a plot into a reservoir. Like its counterpart in West Africa, the inland swamp system impounded water from rainfall, subterranean springs, high water tables, or creeks for soil saturation. Rice cultivation in South Carolina's inland swamps eventually evolved to the point where reserve water could be released on demand for controlled flooding at critical stages of the cropping cycle.[43] The objective of systematic plot irrigation was to drown unwanted weeds and thereby reduce the labor spent weeding. The principle of controlled field flooding was analogous to the one found in the West African mangrove rice system.

Field flooding for irrigation and weed control occurred in a variety of inland swamp settings. For instance, swamps located within reach of streams or creeks often used the landscape gradient for supplemental water delivery. Placement of an embankment at the low end of an undulating terrain kept water on the field while the upper embankment dammed the stream for occasional release. Sluices positioned in each earthen embankment facilitated field drainage and irrigation.[44]

Similar principles sometimes permitted rice planting in coastal marshes near the ocean.[45] Under special circumstances—where a saltwater marsh was located near the terminus of a freshwater stream, for example—rice planting occurred in soils influenced by the Atlantic Ocean. The conversion of a saline marsh to a rice field depended on soil desalination, a result not so quickly achieved under South Carolina's lower annual precipitation (1,100–1,200 mm) as in West Africa, where rainfall in tidal rice-growing areas generally exceeds 1,500 mm per year. Often a creek or stream served the purpose of rinsing salts from the field. Once again the water control system relied on proper placement of embankments and sluices. The lower embankment permanently blocked the inflow of saltwater at high tide, while opening the sluice at low tide enabled water discharge from the plot. A sluice positioned in the upper embankment delivered stream water as needed for desalination, irrigation, and weed control. This type of inland swamp system functioned in the vicinity of the embouchure of the Cooper River, where "rice marshes tempted planters as far down the river as Marshlands [Plantation], nearly within sight of the ocean. Here they had to depend entirely on 'reserve' waters formed by damming up local streams."[46]

The principle of canalizing water for controlled flooding also extended to settings where subterranean springs flowed near the soil surface. Edmund Ravenel described one system that continued to function until the Civil War: "The water here issues from the marl which is about two or three feet below the surface at this spot. This water passes South and is carried under the Santee Canal in a Brick Aqueduct, to be used on the Rice-Fields of Wantoot Plantation."[47]

During the antebellum period another inland swamp system functioned alongside tidewater rice cultivation. While its colonial antecedents remain uncertain, this system flourished where a landscape gradient sloped from rain fed farming to the inner edge of a tidal swamp.[48] Enclosing a tract of land with earthen embankments on high ground created a reservoir for storing rainwater, the system's principal water source. The reservoir fed water by gravity flow to the inland rice field through a sluice gate and canal (figure 7). Excess water flowed out of the plot through a drainage canal and sluice, placed along the lower end of the rice field. The water then drained into a nearby stream, creek, or river.[49] Many techniques of this inland swamp system suggest a West African origin, among them employing a landscape gradient for rice farming, converting a swamp into a reservoir with earthen embankments, and using sluices and canals for water delivery. However, the development of a separate reservoir for water storage perhaps reveals a South Carolina innovation. Only further research can determine whether the creation of a supplementary reservoir for irrigating a swamp

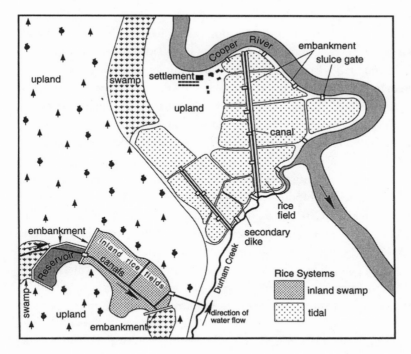

Figure 7
Inland and tidewater rice system, west branch of Cooper River, South Carolina. Adapted from R. Porcher, *A Field Guide to the Bluff Plantation* (New Orleans, 1985).

field is West African, European-American, or the hybridized contribution of both cultures.

By the mid-eighteenth century rice production was steadily shifting from inland swamp systems to tidal river flood plains in South Carolina and into Georgia, just prior to repeal of Georgia's antislavery law in1750.[50] The swelling number of slaves entering South Carolina from the 1730s to the 1770s, plus the fact that these slaves came from West Africa, proved crucial for the transition, and perhaps also to the emergence of the distinctive task labor system that characterized tidal rice cultivation. Some 35,000 slaves were imported into the colony during the first half of the century and over 58,000 between 1750 and 1775, making South Carolina the largest importer of slaves on the North American mainland between 1706 and 1775.[51] The share of slaves brought directly from the West African rice coast grew during these crucial decades of tidewater rice development from 12 percent in the 1730s to 54 percent (1749–65) and then to 64 percent between 1769 and 1774.[52]

One of the earliest references to the existence of the tidal flood-plain system appeared in 1738 with notice of a land sale by William Swinton of Winyah Bay, South Carolina: "that each [field] contains as much River Swamp, as will make two Fields for 20 Negroes, which is overflow'd with fresh Water, every high Tide, and of Consequence not subject to the Droughts."[53] By 1752 rich Carolina planters were converting inland swamps and tidal marshes along Georgia's Savannah and Ogeechee rivers to rice fields, a process actively under way during the 1772 visit by naturalist William Bartram.[54] The shift to tidal production accelerated after the American Revolution, and tidal rice remained the basis of the region's economic prominence until the demise of rice cultivation during the 1920s.[55]

Tidewater cultivation occurred on flood plains along a tidal river where the diurnal variation in sea level resulted in flooding or draining a rice field.[56] Three factors determined the siting of tidewater rice fields: tidal amplitude, saltwater encroachment, and estuary size and shape. A location too near the ocean faced saltwater incursion, while one too far upstream removed a plantation from tidal influence. Like the West African mangrove rice system, a rising tide flooded the fields while a falling tide facilitated field drainage. Along South Carolina rivers tidal pitch generally varied between 1 and 3 feet.[57] These conditions usually prevailed along riverine stretches 10–35 miles upstream from the river's mouth.[58]

Estuary size and shape also proved important for the location of tidewater plantations since these factors affected degree of water mixing and thus salinity. The downstream extension of tidal rice cultivation in South Carolina and Georgia reflected differences in freshwater dynamics between rivers draining the uplands and those flowing inland from the sea. As rivers of piedmont origin deliver freshwater within miles of the coast, tidal cultivation often occurred within a short distance from the ocean. But other coastal rivers are arms of the sea and must reach further inland for freshwater flows (see figure 6). Along such rivers the freshwater stream flow forms a pronounced layer on top of the heavier saltwater, thereby enabling tapping of the former for tidal irrigation.[59] Success under these conditions depended on knowledgeable observation of tidal flows and the manipulation of saltwater-fresh water interactions to achieve high productivity levels in the rice field—skills already belonging to West African tidal rice farmers.

Preparation of a tidal flood plain for rice cultivation followed principles remarkably similar to the mangrove rice system (compare figures 5 and 7). Figure 8 shows the sequence of steps in the conversion of flood plain to rice paddy. The rice field was embanked at sufficient height to prevent tidal spillover, the process leaving a canal adjacent to the embankment. Sluices built into the embankment and field sections operated as valves for flooding and drainage much as they do in Africa's mangrove rice system (figures 5 and 7). The next step involved dividing the area into quarter sections of 10–30 acres, with river water delivered to these sections through secondary ditches. This elaborate system of water control enabled slaves to directly sow rice along the flood plain. Tidewater cultivation required considerable landscape modification and even greater numbers of laborers than the rice-growing systems that first featured prominently in the South Carolina economy. The labor in transforming tidal swamps to rice fields proved staggering, as vividly described by historical archaeologist Leland Ferguson:

These fields are surrounded by more than a mile of earthen dikes or "banks" as they were called. Built by slaves, these banks . . . were taller than a person and up to 15 feet wide. By the turn of the eighteenth century, rice banks on the 12½ mile stretch of the East Branch of Cooper River measured more than 55 miles long and contained more than 6.4 million cubic feet of earth. . . . This means that . . . working in the water and muck with no more than shovels, hoes, and baskets . . . by 1850 Carolina slaves . . . on [tidal] plantations like Middleburg throughout the rice growing district had built a system of banks and canals . . . nearly three times the volume of Cheops, the world's largest pyramid.[60]

The earthen infrastructure continued to make considerable demands on slave labor for maintenance even as it reduced labor spent weeding rice plots.[61] With full water control from an adjacent tidal river, the rice crop could be flooded on demand for irrigation and weeding, and the field renewed annually by alluvial deposits. The historian Lewis Gray underscored the significance of tidal flow for irrigation as well as weeding in explaining the shift from the inland swamp rice system to tidewater cultivation:

Only two flowings were employed [inland swamp] as contrasted with the later period when systematic flowings [tidal] came to be largely employed for destroying weeds, a process which

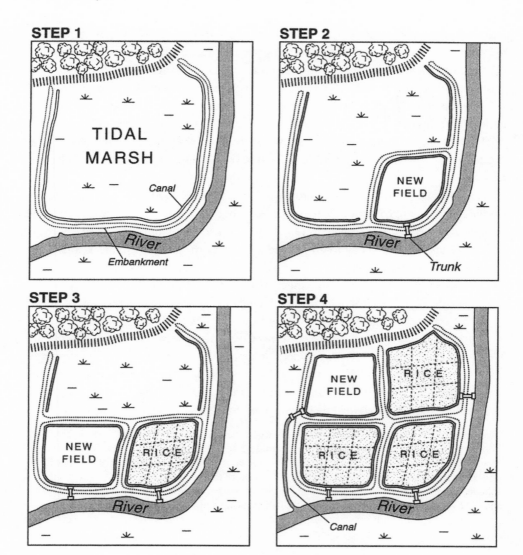

Figure 8
Tidewater swamp conversion, South Carolina. Adapted from S. Hilliard, "Antebellum Tidewater Rice Culture in South Carolina and Georgia," in J. Gibson, ed., *European Settlement and Development in North America* (Toronto, 1978).

is said to have doubled the average area cultivated per laborer. . . . The later introduction of water culture [tidal] consisted in the development of methods making possible a greater degree of reliance than formerly on systematic raising and lowering of the water.[62]

A slave consequently could manage 5 acres instead of the two typically assigned with inland rice cultivation.[63]

The systematic lifting and lowering of water noted by Gray was achieved by the sluices located in the embankment and secondary dikes (figure 7). These crucial devices for water control had assumed the form of hanging floodgates by the late colonial period.[64] As this type of sluice is not traditionally found in West Africa, the hanging gate probably is of European-American origin. Even when this gate replaced earlier forms, the sluices maintained the appellation "trunk" by Carolina planters. The continued use of this term throughout the antebellum period suggests that the technological expertise of Africans indeed proved significant for establishing rice cultivation in the earlier colonial era. During the antebellum period trunks evolved into large floodgates, anchored into the embankment at a level above the usual low tide mark (figure 9). Doors (gates) placed at both ends would swing when pulled up or loosened. The inner doors opened in response to river pressure as the water flowed through the raised outer door and then closed when the tide receded. Field draining reversed the arrangement, with the inner door raised and the outer door allowed to swing while water pressure in the field forced the door open at low tide.[65]

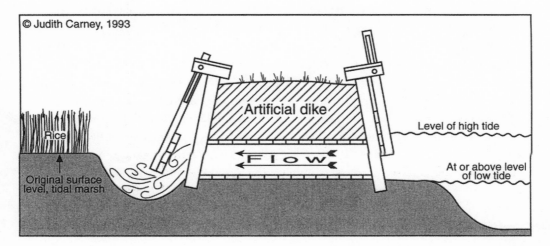

Figure 9
Cross-section of tidewater rice trunk. Adapted from R. Porcher, "Rice Culture in South Carolina: A Brief History, the Role of the Huguenots, and the Preservation of Its Legacy," *Transactions of the Huguenot Society of South Carolina* 92 (1987), p. 6.

Curiosity over the origin of the term "trunk" for sluices or floodgates led planter descendant David Doar to unwittingly stumble on likely technology transfer from West Africa:

For years the origin of this name bothered me. I asked every old planter I knew, but no one could enlighten me. One day a friend of mine who planted on one of the lowest places . . . said to me with a smiling face: "I have solved that little trunk question. In putting down another one, I unearthed the granddaddy of plug trunks made long before I was born." It was simply a hollow cypress log with a large hole from top to bottom. When it was to be stopped up a large plug was put in tightly and it acted on the same principle as a wooden spigot to a beer keg.[66]

The earliest sluice systems in South Carolina looked and functioned exactly like their African counterparts.

Tidewater cultivation led South Carolina to economic prominence in the antebellum era. Its appearance in the colony from the 1730s, and rapid diffusion from mid-century, occurred during a period of escalating slave imports from Africa's rice coast. The evidence presented here concurs with Wood's original claim that Africans tutored planters in developing South Carolina's rice economy. The African experience with planting a whole range of interconnected environments along a landscape gradient likely permitted the sequence of adaptations that marked the growth of the South Carolina rice industry. While the overview of rice cultivation in South Carolina suggests that planters indeed reaped the benefits of a rice farming system perfected by West Africans over millennia, an important question remains: why would West African slaves transfer a sophisticated technology of rice cultivation to the planters when the result harnessed them to brutal toil in malarial swamps?

The answer is perhaps revealed in the appearance of an innovative form of labor organization, the task system, that characterized coastal rice plantations from the mid-eighteenth century. Task labor differed sharply from the more typical "gang" form of work organization, as Gray explains: "Under the task system the slave was assigned a certain amount of work for the day, and after completing the task he could use his time as he pleased." In the gang system "the laborer was compelled to work the entire day."[67] Without overstating the differences in workload between gang and task labor, the task system did set normative limits to the number of hours demanded of slave labor. Such seemingly minor differences, however, could deliver tangible improvements in slave nutrition and health, as Johan Bolzius implied in 1751 with his observation: "If the Negroes are Skilful and industrious, they plant something for themselves after the day's work."[68]

The emergence of the task labor system in South Carolina during the same historical period as accelerating slave imports from West Africa's rice coast and tidewater development is perhaps significant. This form of labor organization may have represented the outcome of negotiation and struggle between master and slave over agronomic knowledge and the labor process. By providing the crucial technological basis for plantation profits, slaves perhaps discovered a mechanism to negotiate improved conditions of bondage. Additional research on the task labor system, whose origins may be African, promises to illuminate the complex relationship between patterns of labor utilization and technical expertise in slave-based plantation systems.[69]

Conclusion

"What skill they displayed and engineering ability they showed when they laid out these thousands of fields and tens of thousands of banks and ditches in order to suit their purpose and attain their ends! As one views this vast hydraulic work, he is amazed to learn that all of this was accomplished in face of seemingly insuperable difficulties by every-day planters who had as tools only the axe, the spade, and the hoe, in the hands of intractable negro men and women, but lately brought from the jungles of Africa."[70] When Doar echoed in 1936 the prevailing view that slaves contributed little besides labor to the evolution of the South Carolina rice economy, no historical research suggested otherwise. While more recent research challenges such assumptions, a bias nonetheless endures against considering the prior rice cultivation experience of African slaves in context of the crop's appearance during the eighteenth century in several areas of the Americas.

In his classic book on global rice cultivation, now in multiple editions, D. H. Grist describes the mangrove system when he writes about empoldering as "a method of restricting floods and thus securing adjacent areas from submergence."[71] He is referring to a type of paddy rice cultivation found in British Guiana (now Guyana) and in the neighboring former Dutch colony of Surinam. On its origins he hypothesizes: "The Dutch are probably responsible for introducing this system into British Guiana in the eighteenth century. Today, all the land developed for paddy cultivation and in the adjacent Dutch colony of Surinam is protected by this means."[72]

This area of northern South America in the late seventeenth through early eighteenth centuries, however, was a plantation society with one of the highest ratios of Africans to Europeans in the Americas (65:1 in Surinam's plantation districts, compared to Jamaica's 10:1).[73] Slave imports continued well into the eighteenth century, with a high percentage originating from the West African rice area discussed by Dutch geographer Dapper in the previous century.[74] Grist's perfunctory treatment of *Oryza glaberrima* in his book (due to the species being "confined to small areas in West Africa [and thus] relatively unimportant") is emblematic of a more pervasive scholarly view toward Africa and its peoples as having contributed little across geographic space besides labor.[75]

This view, however, is giving way as new evidence comes to light from several academic disciplines informed by multi-cultural and cross-cultural perspectives. Recent studies suggest that Europeans and people of European descent can no longer be viewed as the sole masters of technology development and innovation.[76] Yet research on the origins of the rice plantation economy in South Carolina and Georgia displays a lingering Eurocentric bias, granting slaves an initial role in the inland swamp rice system but attributing to planters the crucial technological development of water control that led to tidewater rice cultivation: "Slaves who had experience growing rice in West Africa were probably instrumental in the successful creation of early rice plantations. . . . Some prescient innovators realized that the system would eventually yield diminishing returns and looked for an alternative way to irrigate their crops. The diurnal rising and falling of coastal rivers, caused by the flow and ebb of ocean tides, seemed a likely source of irrigation water. . . . As early as the 1730s, planters noted tidal flow in rivers and, gingerly, began to flow estuarial water over their fields."[77]

The cross-cultural and historical perspective on two important rice-growing regions of the Atlantic basin presented here suggests otherwise. Evidence from the first fifty years of settlement in South Carolina supports the view that technological development and innovation in the rice economy began as an African knowledge system but eventually bore the imprimatur of both African and European influences.[78] By the American Revolution the technological and agronomic heritage of each knowledge system had combined in new ways to shape rice cultivation along the south Atlantic coast of the United States, a process that Paul Richards terms "agrarian creolization."[79] By way of analogy with its linguistic namesake, Richards is referring to the convergence of different knowledge systems (e.g., germ plasm resources and cultivation strategies—in this example, African and Asian germ plasm and African cultivation strategies) and their recombination into new hybridized forms. The outcome of this convergence in South Carolina was a rice production system fashioned from an indigenous African crop that came to bear the distinctive signature of European as well as African culture. Thus, as Africans and Europeans faced each other in new territory under dramatically altered and unequal power relations, the outcome was diffusion, technological innovation, and novel forms of labor organization.

Acknowledgments

This essay originally appeared in *Technology and Culture*,[80] a publication of the Society for the History of Technology. It is reprinted here with the permission of the Johns Hopkins University Press. All figures are by Chase Langford.

Notes

1. *Agriculture of the U.S., United States Census Office 1860* (Washington, 1864).

2. Douglas C. Wilms, "The Development of Rice Culture in 18th Century Georgia," *Southeastern Geographer* 12 (1972): 45–57; Julia Floyd Smith, *Slavery and Plantation Growth in Antebellum Florida, 1821–1860* (Gainesville, 1973) and *Slavery and Rice Culture in Low Country Georgia, 1750–1860* (Knoxville, 1985); Pat Morgan, A Study of Tide Lands and Impoundments within a Three River Delta System—the South Edisto, Ashepoo, and Cumbahee Rivers of South Carolina, M.A. thesis, University of Georgia, 1974; James Clifton, *Life and Labor on Argyle Island* (Savannah, 1978), pp. viii–ix, and "The Rice Industry in Colonial America," *Agricultural History* 55 (1981): 266–283; Charles A. Gresham and Donal D. Hook, "Rice Fields of South Carolina: A Resource Inventory and Management Policy Evaluation," *Coastal Zone Management Journal* 9 (1982): 183–203.

3. Ulrich B. Phillips, *American Negro Slavery* (New York, 1918); A. S. Salley, *The Introduction of Rice Culture into South Carolina* (Columbia, 1919); Ralph Betts Flanders, *Plantation Slavery in Georgia* (Chapel Hill, 1933); David Doar, *Rice and Rice Planting in the South Carolina Low Country* (1936; reprint, Charleston, 1970); Alice Huger Smith, *A Carolina Plantation of the Fifties* (New York, 1936); Norman Hawley, "The Old Plantations in and around the Santee Experimental Forest," *Agricultural History* 23 (1949): 86–91; Duncan Heyward, *Seed from Madagascar* (Chapel Hill, 1937).

4. Converse Clowse, *Economic Beginnings in Colonial South Carolina* (Columbia, 1971); Peter Wood, *Black Majority* (New York, 1974), pp. 57–64.

5. Daniel Littlefleld, *Rice and Slaves* (Baton Rouge, 1981). Betty Wood, *Slavery in Colonial Georgia* (Athens, 1984), p. 103, indicates a similar trend for Georgia, noting that three-fourths of the slaves shipped there during the critical period of tidewater rice expansion (1766–1771) originated from West Africa's rice coast.

6. Roland Porteres, "Primary Cradles of Agriculture in the African Continent," in *Papers in African Prehistory*, ed. J. Fage and R. Oliver (Cambridge, 1970), pp. 43–58; Jack Harlan, J. De Wet, and A. Stemler, *Origins of African Plant Domestication* (Chicago, 1976); R. Charbolin, "Rice in West Africa," in *Food Crops of the Lowland Tropics*, ed. C. Leakey and J. Wills (Oxford, 1977), pp. 7–25.

7. Charles Kovacik and John Winberry, *South Carolina: The Making of a Landscape* (Boulder, 1987); Timothy Silver, A *New Face on the Countryside* (New York, 1990).

8. Judith Carney, The Social History of Gambian Rice Production: An Analysis of Food Security Strategies, Ph.D. dissertation, University of California, Berkeley, 1986.

9. Ibid., p. 23; George Brooks, *Landlords and Strangers: Ecology, Society, and Trade in Western Africa, 1000–1630* (Boulder, 1993), pp. 9–13.

10. F. R. Moorman and W. J. Veldkamp, "Land and Rice in Africa: Constraints and Potentials," in *Rice in Africa*, ed. I. Buddenhagen and J. Persely (London, 1978); West African Rice Development Association, *Types of Rice Cultivation in West Africa*, Occasional Paper no. 2 (Monrovia, 1980); Carney, "Social History of Gambian Rice."

11. H. A. R. Gibb, *Ibn Battuta: Travels in Asia and Africa, 1325–1354* (London, 1969); Graham Connah, *African Civilizations* (New York, 1987).

12. Tadeusz Lewicki, *West African Food in the Middle Ages* (Cambridge, 1974).

13. See, e.g., Paul Pélissier, *Les paysans du Sénégal: Les civilisations agraires du Cayor à la Casamance* (Saim-Yrieix, 1966), pp. 711–712. Pélissier quotes Azurara's 1446 observation in the vicinity of the Gambia River: ". . . they found there a river of great expanse, which they entered with their caravels. . . . [Some of the men] landed . . . and following for some distance said they found the country covered with a great deal of territory sown to rice." Pélissier also quotes Eustache de la Fosse on his 1479–80 visit to the littoral rivers of southern [Casamance] Senegal and Guinea-Bissau: "We had . . . good rice and good milk. . . . I asked our captain [pilot] the origin of this good rice. . . . He told me . . . they had arrived at the Ydolles Islands and found that the Blacks abounded in the goods. At their market place they had several large mounds of rice . . . and all was brought to the ships." (author's translations) See also Gomes Eannes de Azurara, *The Chronicle of the Discovery and Conquest of Guinea*, vol. 2 (London, 1899); G. R. Crone, *The Voyages of Cadamosto* (London, 1937); A. Donelha, *An Account of Sierra Leone and the Rivers of Guinea and Cape Verde* (Lisbon, 1977).

14. Valentim Fernandes, *Description de la Côte Occidentals d'Afrique* (Bissau, 1951); Walter Rodney, *A History of the Upper Guinea Coast, 1545–1800* (New York, 1970).

15. For more detailed discussion of these early European commentaries on rice cultivation along the West African Atlantic coast, see Judith Carney, "From Hands to Tutors: African Expertise in the South Carolina Rice Economy," *Agricultural History* 67 (1993): 1–30.

16. Translation and excerpt drawn from Olfert Dapper, *New Description of Africa*, by Paul Richards, in "Culture and Community Values in the Selection and Maintenance of African Rice," paper presented at Conference on Intellectual Property Rights and Indigenous Knowledge, Lake Tahoe, 1993. On Dapper, see also Adam Jones, *From Slaves to Palm Kernels* (Wiesbaden, 1983) and "Decompiling Dapper: A Preliminary Search for Evidence," *History in Africa* 17 (1990), 171–209.

17. M. Adanson, A *Voyage to Senegal, the Isle of Gorée and the River Gambia* (London, 1759); Francis Moore, *Travels into the Inland Parts of Africa* (London, 1738), G. Mollien, *Travels in Africa* (London, 1820); Mungo Park, *Travels into the Interior of Africa* (1799; reprint London, 1954); Pélissier, *Les paysans du Sénégal*.

18. Rodney Thomas Winterbottom, An *Account of the Native Africans in the Neighbourhood of Sierra Leone* (London, 1803); Rene Caillié, *Travels through Central Africa to Timbuctoo, and across the Great Desert, to Morocco, Performed in the Years 1824–1828* (London, 1830).

19. Pierre Viguier, *La risiculture indigene au Soudan Français* (Paris, 1939); Littlefield, *Rice and Slaves*; Paul Richards, *Indigenous Agricultural Revolution* (London, 1985); Richards, *Coping with Hunger* (London, 1986).

20. W. Andriesse and L. O. Fresco, "A Characterization of Rice-Growing Environments in West Africa," *Agriculture, Ecosystems and Environment* 33 (1991): 377–395.

21. The discussion here is based on my own fieldwork in Senegambia over a ten-year period, as well as on research of even longer duration in Casamance, Senegal, by Olga F. Linares and in Sierra Leone by Paul Richards. See, e.g., Olga F. Linares, "From Tidal Swamp to Inland Valley: On the Social Organization of Wet Rice Cultivation among the Diola of Senegal," *Africa 5* (1981): 557–594; Linares, *Power, Prayer and Production* (New York, 1992); Richards, *Indigenous Agricultural Revolution* and *Coping with Hunger*.

22. F. R. Moorman and N. Van Breeman, *Rice: Soil, Water, Land* (Los Banos, 1978).

23. Richards, *Coping with Hunger;* Andriesse and Fresco, "Characterization of Rice-Growing Environments."

24. Richards, "Culture and Community Values." See also C. Fyfe, *A History of Sierra Leone* (Oxford, 1962), p. 4.

25. Caillié, *Travels through Central Africa*, p. 162.

26. Francesca Bray, "Patterns of Evolution in Rice Growing Societies," *Journal of Peasant Societies* 11 (1983): 3–33.

27. Food and Agriculture Organization, *Rice Mission Report to the Gambia* (Rome, 1983).

28. Littlefield, *Rice and Slaves*; Joyce Chaplin, *An Anxious Pursuit: Agricultural Innovation and Modernity in the Lower South, 1730–1815* (Chapel Hill, 1993).

29. See, for instance, Littlefield, *Rice and Slaves*, p. 86.

30. Pélissier, *Les paysans du Sénégal*; Linares, "From Tidal Swamp" and *Power, Prayer and Production*.

31. Heyward, *Seed from Madagascar*; Doar, *Rice and Rice Planting*; Chaplin, *An Anxious Pursuit*.

32. Rodney, *History of the Upper Guinea Coast*.

33. Linares, "From Tidal Swamp" and *Power, Prayer and Production*.

34. Albert Virgil House, *Planter Management and Capitalism in Ante-Bellum Georgia* (New York, 1954); James Clifton, "Golden Grains of White: Rice Planting on the Lower Cape Fear," *North Carolina Historical Review* 50 (1973): 365–393; Clifton, "Rice Industry."

35. Wood, *Black Majority*, pp. 25–26, 36, 143–145. Eugene Sirmans (*Colonial South Carolina*, Chapel Hill, 1986, p. 24) claims that South Carolina showed a preference for slave rather than indentured labor from the earliest period of settlement. He attributes this anomaly before development of a plantation economy to the role of African labor. Perhaps the argument merits extension to the value of African knowledge of rice cultivation.

36. Wood, *Black Majority*, pp. 57–58.

37. Clifton, "Rice Industry," p. 269.

38. Wood, *Black Majority*, p. 55.

39. Ibid., pp. 57–58. No official document mentions rice planting before 1690 even though it was considered a suitable crop for Carolina by Spanish, French, and English officials prior to the colony's settlement. See Clifton, "Rice Industry," p. 270, for the earliest indirect evidence for rice growing in South Carolina: several English runaways to St. Augustine, Florida, claimed in 1674 that "some rice . . . grown on the soil was shipped to Barbados."

40. Doar, *Rice and Rice Planting*; Kovacik and Winberry, *South Carolina*; Judith Carney and Richard Porcher, "Geographies of the Past: Rice, Slaves and Technological Transfer in South Carolina," *Southeastern Geographer* 33 (1993): 127–147.

41. John S. Otto, *The Southern Frontiers, 1607–1860* (New York, 1989); Wood, *Black Majority*, pp. 30–32, 105–114; Terry Jordan, *Trails to Texas: Southern Roots of Western Cattle Ranching* (Lincoln, 1981), pp. 14, 29, 33.

42. On numbers of imported slaves, see Clowse, *Economic Beginnings*, p. 252; Wood, *Black Majority*, pp. 143–145; Peter Coclanis, *The Shadow of a Dream* (New York, 1989), p. 64. For descriptions and periodization of the rain-fed and inland swamp systems, see Thomas Nairne, "A Letter from South Carolina," in *Selling a New World*, ed. J. Greene (1710; reprint, Columbia, 1989), pp. 33–73; Lewis Gray, *History of Agriculture in the Southern U.S. to 1860* (Gloucester, 1958), vol. 1, p. 279; Clifton, *Life and Labor*; Clarence Ver Steeg, *Origins of a Southern Mosaic* (Athens, 1984); Otto, *The Southern Frontiers*.

43. Heyward, *Seed from Madagascar*; Sam B. Hilliard, "Antebellum Tidewater Rice Culture in South Carolina and Georgia," in *European Settlement and Development in North America*, ed. J. Gibson (Toronto, 1978); Richard Porcher, "Rice Culture in South Carolina: A Brief History, the Role of the Huguenots, and the Preservation of Its Legacy," *Transactions of the Huguenot Society of South Carolina* 92 (1987): 11–22; David Whitten, "American Rice Cultivation, 1680–1980: A Tercentenary Critique," *Southern Studies* 21 (1982): 5–26.

44. Clifton, "Rice Industry," p. 275.

45. Hawley, "The Old Plantations."

46. John B. Irving, *A Day on the Cooper River* (Charleston, 1969), p. 154. The remains of a similar system can be seen behind Murphy Island in the Santee Delta where an extensive salt marsh is located near a freshwater stream, and at Drayton Hall on the Ashley River.

47. Edmund Ravenel, "The Limestone Springs of St. John's, and their probable Availability for increasing the quantity of Fresh Water in Cooper River," *Proceedings of the Elliott Society of Science and Art, of Charleston, South Carolina*, 2 (October 1860), p. 29.

48. Hilliard ("Antebellum Tidewater Rice Culture," p. 99) points out that during much of the eighteenth century both inland swamp fields with reservoirs and tidewater rice existed simultaneously and that freshwater reservoirs were common even on plantations situated within or near the tidal zone: "In many cases there must have been a blending of the two types of irrigation, for Solon Robinson observed a tidewater planter on the Cooper River who had 'ponds of fresh water covering 100 acres of upland, which are held in reserve to water the rice fields when the river is too salt.'"

49. Porcher, "Rice Culture in South Carolina."

50. Wilms, "The Development of Rice Culture"; Smith, *Slavery and Rice Culture*.

51. David Richardson, "The British Slave Trade to Colonial South Carolina," *Slavery and Abolition* 12 (1991): 125–172, esp. pp. 127–128.

52. Ibid., pp. 135–136.

53. *South Carolina Gazette*, January 19, 1738. Clifton ("Rice Industry," pp. 275–276) observes notices of tidal swamps for sale first appearing during the 1730s in the *South Carolina Gazette*—1731 for the Cape Fear River and 1737 for the Black River in South Carolina.

54. Wilms, "The Development of Rice Culture," p. 49.

55. Clifton, "Rice Industry," p. 276.

56. Hilliard, "Antebellum Tidewater Rice Culture," p. 100.

57. Chaplin, *An Anxious Pursuit*.

58. John Drayton, *View of South Carolina* (1802; reprint, Spartanburg, 1972), p. 36; Chaplin, *An Anxious Pursuit*, p. 231.

59. Hilliard, "Antebellum Tidewater Rice Culture."

60. Leland Ferguson, *Uncommon Ground: Archaeology and Early African America, 1650–1800* (Washington, 1992), pp. xxiv–xxv, 147.

61. This herculean toil is likely responsible for the gradual demise of tidewater cultivation after emancipation and for freedmen's avoidance of "mud work" as hired labor on tidal plantations even as they continued to grow rice on their own as a cash crop in some inland swamps of South Carolina until the 1930s; see, e.g., Amelia Wallace Vernon, *African Americans at Mars Bluff, South Carolina* (Baton Rouge, 1993). Vernon documents the survival of rice cultivation among African-Americans as a response to blacks' restricted access to farmland following Reconstruction. Cultivating rice as a cash crop on unclaimed swamps represented an important form of resistance to the prevailing exploitative wage labor and sharecropping relations that emerged. Planting the crop of their forebears consequently nurtured freedmen's dreams of independent farming, the failed promise of Reconstruction.

62. Gray, *History of Agriculture*, p. 281.

63. R. F. W. Allston, "Essay on Sea Coast Crops," *De Bow's Review* 16 (1854): 589–615; James Glen, "A Description of South Carolina: Containing Many Curious and Interesting Particulars Relating to the Civil, Natural and Commercial History of That Colony," in *Colonial South Carolina*, ed. C. Milling (Columbia, 1951), p. 15; Clifton, "Rice Industry," p. 275; Whitten, "American Rice Cultivation," pp. 9–15.

64. Richard Porcher, *A Field Guide to the Bluff Plantation* (New Orleans, 1985), pp. 26–27.

65. House, *Planter Management and Capitalism*, p. 25.

66. Doar, *Rice and Rice Planting*, p. 12

67. Gray, *History of Agriculture*, pp. 550–551; Philip Morgan, "Work and Culture: The Task System and the World of Low Country Blacks, 1700 to 1880," *William & Mary Quarterly* 39 (1982): 563–599; Smith, *Slavery and Rice Culture*, p. 61.

68. Bolzius, quoted in Morgan, "Work and Culture," p. 565. The entire document appears in "Johan Bolzius Answers a Questionnaire on Carolina and Georgia," ed. K. Loewald, B. Starika, and P. Taylor, *William & Mary Quarterly* 14 (1957): 218–261.

69. Carney, "From Hands to Tutors," pp. 26–28.

70. Doar, *Rice and Rice Planting*, p. 8.

71. D. H. Grist, *Rice*, fourth edition (London, 1968), p. 45. This comment also suggests the need for additional research on how the Dutch used poldering techniques for agriculture.

72. Ibid.

73. See, e.g., Richard Price and Sally Price, *Stedman's Surinam: Life in an Eighteenth-Century Slave Society* (Baltimore, 1992), pp. xii, 208–219. Fugitive slave communities in eighteenth-century Surinam widely cultivated rice, experienced abundant harvests, transferred the crop between

communities, and even took their eponyms from rice (example: Reisse Condre, meaning "from the quantity of rice it afforded").

74. Ibid.

75. Grist, *Rice*, p. 56.

76. Doar (*Rice and Rice Planting*, p. 20) once again illustrates this Eurocentric bias against African as well as Asian knowledge systems in his 1936 book: "It is one hundred and fifty-seven years [sic] since the introduction of rice into Carolina, and there are grounds for supposing that our people have accomplished more during that period, in the cultivation and preparation of this grain, than has been done by any Asiatic nations, who have been conversant with its growth for many centuries." For recent work that counters the attribution of European political-economic hegemony to technological superiority over other cultures, see Michael Adas, *Machines as the Measure of Men* (Ithaca, 1989); Jim Blaut, "On the Significance of 1492," *Political Geography* 11 (1992): 355–385.

77. This is the point of view set forth by Chaplin in *An Anxious Pursuit*. The quotations are from pp. 228, 231, and 232.

78. On possible diffusion of Asian rice techniques and technologies into South Carolina via the botanical gardens and scholarly societies that proliferated with European political-economic expansion across the globe, see Chaplin, *An Anxious Pursuit*, pp. 147–150.

79. Richards, "Culture and Community Values," p. 2. Several decades ago, Melville Herskovits drew attention to the significance of the idea of cultural "creolization" for research: "The problem . . . was the manner in which elements of European, African and . . . American Indian cultures had exerted mutual influences." Melville Herskovits, in *The New World Negro: Selected Papers in Afroamerican Studies*, ed. F. Herskovits (Bloomington, 1966), pp. 36–37.

80. *Technology and Culture* 37 (1996), January: 5–35.

"To Collect Proof of Colored Talent and Ingenuity": African-American Invention and Innovation, 1619–1930

Portia James

We are just beginning to uncover the history of African-American technology, just beginning to understand its relation to the full story of American invention and innovation, and now finally seeing that these histories are intertwined from the start. Enslaved Africans, carried against their will to this country, inevitably brought with them a store of technological knowledge, ways of doing things, that they applied to the daily processes of life in this unfamiliar environment. Black people have been here since the earliest times and, despite the perception that they were merely a source of ignorant labor, they had an immediate impact on the technologies around them—applying traditional skills, creating some anew, and altering the nature of still other practices. In fact, the New World was a place where many different cultures and technologies met. To say that we were culturally diverse from the first is also to say that our technologies were, too.

Increasingly, we discover the African roots of many early craft activities. Theresa Singleton, who has studied archeological sites in Berkeley County, South Carolina, points out that "West African culture was thoroughly woven into the daily lives of South Carolinians in the colonial era."[1] In South Carolina, slave houses were built with African building technology, while in the Chesapeake region excavated clay pipes reveal the decorative techniques of West African pottery.[2] The study of early technology not only illustrates how African craft skills were incorporated into American technology, but how Africans were themselves incorporated into early American society. The clay pipes and pottery of the Chesapeake, for example, tell us something about the changing status of seventeenth-century blacks in America. It was not uncommon then for black slaves and white indentured servants to work together alongside their masters, and for black and white artisans to exchange craft techniques and trade secrets with each other. But by the beginning of the eighteenth century, as the previously varied forms of servitude congealed into black and white categories of slave or not slave, evidence of collaborative work disappears.[3]

Another example of the merging of different technological traditions is in the introduction, use, and development of dugout canoes in the Chesapeake region. Common in both the Caribbean, where Native American and later black boatmen used them for fishing, and in West Africa, where people depended on a vast network of rivers for transportation, dugout canoes were widely employed and modified according to circumstance. When West Indian immigrants began settling in the Chesapeake, the Africans among them continued their tradition of building boats from a single log, sometimes adding planks on each side to increase load-bearing capacity. John Vlach, who has studied the appearance of the pirogue in colonial America, has suggested that these West Indian blacks were the first to construct that type of boat in the Chesapeake in the early 1700s. And just as one can see evidence of the connections between African and European traditions in colonial era pottery making, varieties of log canoe making in the Virginia and Chesapeake areas reflect the same interrelatedness.[4]

Africans brought from the West Indies to the Carolinas in the seventeenth century came with rice growing techniques as part of their cultural baggage, and it proved an important resource when English colonists—with little experience or knowledge of that crop—sought to exploit the marshy soils of the Carolina lowlands. These people, having originated in what was called the Rice Coast of West Africa, used their own traditional planting techniques and also knew the most efficient ways of hulling and cleaning rice. African-American mortars and pestles made expressly for hulling rice often show African designs, as do such other implements as the wide, shallow baskets that women wove from local grasses and then used to separate rice from chaff.[5] This range of transplanted techniques provided the foundation for the principal agricultural crop of the Carolinas, and the wealth that flowed from it.

Knowledge, experience, and skill lie at the heart of creative technical activity, whether simple or complex. And it is clear that African-Americans possessed all the necessary ingredients for making improvements in the technologies they used. But before 1865, and the passage of the thirteenth and fourteenth amendments to the Constitution, the impediments to an actual career of inventing, to the enjoyment of any economic rewards from their inventions—much less the recognition other inventors of this country received—kept most African-Americans from a public recording of their talents. Enslaved blacks were prohibited by law from the patent system because they were constitutionally defined as noncitizens. Still, there were frequent reports of slaves who developed improved ways of doing things, not surprising since so many of them possessed domestic, agricultural, and mechanical skills.

The case of "Ned" nicely illustrates what happened when a slave came up with an idea that promised financial gain. An enslaved mechanic in Pike County, Mississippi,

Ned had devised a cotton scraper that local planters claimed would enable its users to do twice the work with half the horsepower. Consequently, Ned's owner, a prominent local planter named Oscar J. E. Stuart, wrote in 1857 to the Secretary of the Interior, Jacob Thompson, seeking letters patent for himself as Ned's owner, and therefore the owner as well of Ned's invention. And Stuart purposely sought out Thompson, as "a Mississippian and Southern man," to help him with his claim, reminding him that in the tradition of Southern law, "the master is the owner of the fruits of the labor of the slave both intilectual [*sic*], and manual."[6] Concerned that Patent Office officials might think of awarding the patent to Ned himself, Stuart argued that in such a case "the value of the slave to his master is excluded, and the equal protection and benefit of government to all citizens . . . is subverted."[7]

In another letter, this time to the Commissioner of Patents, Joseph Holt, Stuart again applied for the rights to Ned's invention, and he included with his letter an "affidavit" signed by Ned to the effect that the black man had invented the machine and was indeed the slave of Stuart. But Commissioner Holt denied Stuart's claim, on the grounds that noncitizens could not apply for patent rights in the United States, an opinion in which the Attorney General concurred. Stuart's enraged response fully reveals the context in which people like Ned found themselves:

I was never such an unmitigated fool which is the implication of the act of the Commissioner as to imagine that a slave could obtain a patent for a useful invention when under the laws, it is a question . . . whether the master who has a property alike in the fruits of the mind and labor of the hands of his slave whose automaton in legal contemplation he is . . . can obtain a patent when the invention is made by him. . . . [The Commissioner] has made up a hypothetical case as though the slave Ned had petitioned for a patent for the invention & decided he could not entertain it. . . . For if [Ned] has ever had any correspondence with [the Patent] bureau upon the subject I am ignorant of it, and for such impertinence, you know according to our Southern usage, I would correct him.[8]

Owners like Stuart might have expected their slaves to behave as automata, but they wanted the benefits of their brains as well. So when the Confederate States established their own patent act, one of its central provisions gave masters the rights to the inventions of their slaves.

Yet, in spite of law and custom, restrictions and hostility, free and unfree African-Americans in the decades before the Civil War brought their ideas for improvements in technology to a wide range of economic activity. Many of their inventions sprang directly out of the craft pursuits in which they were engaged, which is true of most inventions, and also why it is important to recognize the link between skill and creativity. We can see how that relation played out as Northern port cities—more cosmopolitan and at the same time more anonymous—proved a magnet for free blacks and escaped slaves seeking employment, and the maritime trades in those places became the major source of livelihood for them. For instance, at least one-fifth of the merchant seamen in Philadelphia in 1796 were free blacks, and by 1846, according to the *National Anti-Slavery Standard*, there were 6,000 African-American seamen in port cities around the country. Boston and New York became important centers of maritime employment for black men, as did the whaling ports of Nantucket and New Bedford, Massachusetts, where the population of African-Americans doubled in the 1830s.[9] Most of those men labored as ordinary seamen, but a few rose to command. The whaling ship *Loper*, for example, was commanded by a black captain and black navigators, and was manned by an almost entirely black crew.[10] Out of their experience in these trades, they also became innovative. James Forten (1766–1842) is a good example of the case. During the American Revolution, Forten, age 14, signed on the privateer *Royal Louis* where he served as a powder boy, alongside twenty other blacks. His ship was captured and he spent time in a British prison hulk before being released in a general prisoner exchange. On his return to Philadelphia, he apprenticed to the sailmaker Robert Bridges, who had previously employed Forten's father. Obviously skilled, Forten was made foreman of the sail loft in 1786, and twelve years later became owner of the firm. At some point in that career, he invented, but never patented, some kind of sail handling device that reputedly helped make his business financially more successful.[11] With his profits, Forten funded antislavery activities and became a prominent abolitionist pamphleteer and spokesman. His history not only reminds one of Frederick Douglass, who learned the caulker's trade in Baltimore shipyards, before himself becoming a celebrated antislavery orator and publisher—but also of the vital connection Douglass sought to establish between the mastery of craft skill and a consequent manly independence of thought and action.

Lewis Temple (1800–1854) was another African-American in the maritime trades who turned his skills to invention. Born in Richmond, Virginia, Temple migrated to New Bedford and by 1836 had set up shop on Coffin's Wharf as a blacksmith to the whaling trade. He moved again, in 1845, to the Walnut Street Wharf, and it was in that shop that he developed the Temple toggle harpoon, a modification to the ordinary harpoon to prevent it from being pulled loose as the whale struggled. There were many ideas for improved harpoons floating around the whaling ports, but what made Temple's different, and successful, was that he incorporated in it a wooden shear pin that broke as the whale thrashed, thus releasing the toggle at right angles to the shaft and so making it fast. According to Sidney Kaplan, his biographer, Temple's device became "the universal whale iron," and remained so for a long time.[12] Like Forten, however, Temple never patented his improvement, even though as free men they had the legal right. Both class and color militated against patent applications by people like these; costs were involved in the application itself, a model and drawings to exemplify the improvement being claimed were also an expense, as were the services of a patent attorney to speed the process. And almost always, there was the reality of racial prejudice. The abolitionist and colonization advocate Martin R. Delaney (1812–1885), for example, developed a device that would assist railroad locomotives in descending and ascending inclined planes, and in 1852 traveled to New York to get a patent for it. He engaged the services of a patent attorney, but out of ignorance, or ill will, or both, Delaney was advised by his counselor that blacks were not considered citizens by the Patent Office, and that he should give up his application.[13]

These same factors, of course, also made it difficult for African-Americans to gain access to apprenticeship programs. Frederick Douglass was driven from his craft by the utter hostility of white ship caulkers, for example, and free blacks—North or South—faced restrictions on their freedom of movement, laws disallowing their testimony in court, the threat of disenfranchisement, or even enslavement for minor offenses, and often what amounted to a tacit boycott of their practice.[14] But even though most free blacks worked as laborers or domestics, they could be found in virtually every craft, engaged as potters, tailors, blacksmiths, coopers, carpenters, and even silver and goldsmiths. Apprenticeships in the best paying crafts were least accessible, and the reverse was true, too. Tailoring was a relatively easy craft for African-Americans to enter, whether male or female. Thomas L. Jennings (1791–1859), for example, learned tailoring and dry cleaning and carried on that business in New York City, where in 1821, at the age of thirty, he received a patent for an improvement in dry cleaning processes—perhaps the first black person in this country to have received one.[15] And as was the case of so many others like him, Jennings used the income from his business to support his abolitionist activities. Elizabeth Keckley (1818–1907) took the earnings from her skills as a seamstress to buy her freedom from slavery, and then developed a system for cutting and fitting dresses that she taught to other dressmakers in Washington, D.C.[16]

Carpentry was another common trade for African-American skilled workers, and furniture makers formed a kind of elite within this group, enjoying higher wages and status. Henry Boyd (1802–1886), a Cincinnati furniture maker, derived much of his success from his invention of the Boyd bedstead—a wooden bed frame designed so that its wooden rails could be screwed into the headboard and the footboard simultaneously, thus creating a stronger frame. Boyd's personal history suggests the kind of determination it took for a black person to make a success in business. In the first place, he had been apprenticed to a cabinet maker as a slave boy, but worked nights in a salt-processing plant to gain the money he needed to buy his freedom. Once in Cincinnati, he hired himself out as a laborer until he was able to form a house building partnership with a white carpenter, and that business provided the funds with which to buy his brother and sister out of slavery. By 1836 he owned his own bedstead factory, where he employed from up to fifty workers and used steam-powered machinery. He was burned out by arsonists twice, and rebuilt his business both times, selling most of his beds in the south and southwest.[17] Boyd did not patent his improvement, instead resorting to identifying his beds with a stamp that bore his name—a mark that now gives special value to collectors of the region's furniture.

John Parker (1827–1900) was another slave who bought his freedom with craft skill. Parker began working in the iron foundries of Mobile, Alabama, where he was apprenticed to an iron molder, but took on extra work to earn the money with which he purchased himself from his master. He moved to Ohio in 1850, established a foundry, became active in the Underground Railroad movement, and invented a tobacco screw press, among other devices.[18] In fact, a great many enslaved blacks labored in Southern industrial establishments, especially iron working but also in shipyards, mines, and cotton mills—and Parker was far from the only one to use his skills to hire out for the extra money that bought freedom.

In a country that placed such a cultural value on inventiveness, African-Americans naturally sought to prove their own worthiness by technical accomplishment, and there is a long history of explicit efforts to establish that connection. Perhaps the most celebrated was Benjamin Banneker's correspondence with Thomas Jefferson. In his *Notes on the State of Virginia*, published in 1781, Jefferson had argued the intellectual inferiority of the African, saying "one could scarcely be found capable of tracing and comprehending the investigations of Euclid."[19] Banneker, an ardent mathematician and astronomer, used his astronomical calculations to produce an almanac in 1791, and sent a copy to Jefferson by way of refuting that argument with the observation, "we have long been considered rather as brutish than human, and scarcely capable of mental endowments."[20] Nineteenth-century African-American activists remained just as aware of the political value of inventors, who were held up as explicit examples in the attack on claims of mental inferiority. To that end, the country's foremost abolitionist newspaper, *Liberator*, published a notice in its issue of September 6, 1834, requesting information on "colored inventors of any art, machine, manufacture, or composition of matter." His aim, the editor said, was "to collect proof of colored talent and ingenuity," but also "to aid colored inventors in obtaining their patents for valuable inventions."

These efforts to publicize the potential of black inventiveness became a staple of public assemblies, too. At the 1858 Convention of the Colored Citizens of Massachusetts, William C. Nell offered the following resolution:

Resolved, That we rejoice in the presence here today of Mr. Alexander [Aaron] Roberts of Philadelphia, the inventor of a machine for use at fires, which promises to be one of utility in their extinction, as also for preserving human life.

Resolved, That we would also direct attention to the new railway, by which space is economized, the use of horses obviated, and at the same time propelled by steam power; said railway being the invention of a colored man, William Deitz, of Albany, N.Y.

Resolved, That we commend these colored American Inventors and their inventions to the favorable attention of every lover of science and well-wisher of Humanity.[21]

And in an era or industrial exhibitions, African-Americans organized fairs and institutes to promote the accomplishments of skilled craftspeople and inventors. One of the earliest, The Colored American Institute for the Promotion of the Mechanic Arts and Sciences, was opened in Philadelphia on April 12, 1851. Its aim was to exhibit the work of black mechanics, artisans, and inventors, thus bringing them to the attention of clients, but also to exhibit proofs of black talent. Abolitionist newspapers lauded such fairs as glimpses of the kinds of achievements possible in the race, once freed from the constraints of slavery.

The reality is that most inventions submitted to the Patent Office in the nineteenth century consisted of minor improvements to the ordinary, workaday devices used by a rural people. There is no end to the number of patents for washing machines, butter churns, and farming tools, but that is because such things were at the heart of ordinary life, and because so many Americans imagined it possible to improve what they knew best. We see exactly the same eager ingenuity as African-Americans turned their hands to inventiveness. George Peake (1722–1827) settled in northeastern Ohio in 1809, and invented a hand-operated mill for grinding wheat into flour. Henry Blair, a Montgomery Country, Maryland black man, received a patent for a corn planter in 1834 and another for a cotton planter in 1836.[22] And there are many others, known only anecdotally, who are said to have developed cotton-cleaning machines, broommaking machines, systems for curing tobacco, and so on. The number of such stories suggests considerable creative vitality in the countryside, which can only be guessed at without the detailed historical information of patent records.

But if most inventions required little theory and less science, some called for considerable technical training, and even in the period before the Civil War there are examples of that level of inventiveness. Norbert Rillieux (1806–1894) provides the best known case, even if untypical. Born in New Orleans, Rillieux was the son of Constance Vivant, a quadroon, and Vincent Rillieux, a white engineer. His family, wealthy enough and with a French heritage, sent him to L'Ecole Centrale in Paris to receive a technical education. He later taught mechanical engineering at the school, and soon began to devote himself to questions of thermodynamics and the applications of steam power. In 1830 Rillieux conceived the multiple effect vacuum evaporation system, which he later patented. But, unable to persuade anyone in Paris to invest in his ideas, he returned to New Orleans with the hope of interesting someone there in the application of his improved methods to the refining of cane sugar. He obtained his first patent in 1843, but it was not until 1845 that he was able to put his evaporation techniques into successful operation. The system proved so efficient that planters could cover the cost of the machinery with the extra profits from the first crop of sugar cane processed by the new technology. The Rillieux method revolutionized the sugar industry by dramatically reducing the cost of producing refined sugar, and so making white sugar widely and cheaply available.[23] He became wealthy from this invention, but as New Orleans became a less hospitable place for free blacks, enacting new discriminatory legislation, Rillieux returned to France and died there.

The end of the Civil War and the passage of the Thirteenth and Fourteenth Amendments meant that all black inventors now had the right to apply for patents. The result over the next few decades was a virtual explosion of patented inventions by black mechanics, blacksmiths, domestic workers, and farm laborers—many of them ex-slaves. By 1895 the U.S. Patent Office was able to advertise a special exhibit of inventions patented by black inventors.[24]

The list of new inventions patented by blacks after the Civil War reveals what kinds of occupations they held and in which sectors of the labor force they were concentrated. Agricultural implements, devices for easing domestic chores, and devices related to the railroad industry were common subjects for black inventors. Some patented inventions developed in the course of operating businesses like barber shops, restaurants, and tailoring shops. Joseph Lee, who had highly successful catering and restaurant businesses, considered himself a "bread specialist," and invented bread-making and bread-crumbing machines.[25] Alexander Ashbourn, of Oakland, California, received several patents in the 1870s related to food preparation, including patents for products derived from coconuts. He later moved to Boston and then to Philadelphia, where he began manufacturing and selling such goods as tooth powder, ink, vinegar, and soap—all made from coconuts. Henry A. Bowman, of Worcester, Massachusetts, began a business that made awnings, tents, canvas covers, and flags—the latter by an improved method that he patented. R. N. Hyde left Culpepper, Virginia, for Des Moines, Iowa, where he established a custodial service in 1880. The experience led him to invent an electric carpet cleaning machine and to develop a number of cleaning compounds. He sold his profitable business in 1905 and became prominent in Republican Party politics.[26] And some black women inventors of this period have been identified, too. The Patent Office awarded a patent to Julia Hammonds for a knitting device, and one to Sarah E. Good for a folding cabinet.[27]

Men who had been blacksmiths and mechanics during slavery, and who were able afterwards to accumulate enough start-up capital, opened mechanical and iron-working businesses. William Powell started his own firm, the Standard Repair Shop, in Cass County, Michigan, in a community settled by escaped slaves, and developed a reputation for his inventiveness.[28] Frank J. Ferrel, a skilled machinist and an ardent labor unionist, was a delegate and organizer for the Knights of Labor, one of the earliest national labor unions. He patented several valves, a steam trap, as well as an apparatus for melting snow, and these inventions became the basis for a manufacturing company that he established in New York City.[29] Joseph H. Dickinson worked at the Clough & Warren Organ Company in Detroit, Michigan as a young man. Then in 1882 formed a partnership with his father-in-law to create the Dickinson-Gould Organ Company, in Lexington, Michigan, making parlor and chapel organs. The company sent a large chapel organ to the New Orleans Exposition of 1884, as part of an exhibit demonstrating the accomplishments of black people, and in the 1890s Dickinson received patents for several improvements in reed organs.[30]

Two important points emerge from these various examples. One is that even in the face of poverty and discrimination, black people were as caught up in the appeal of invention as other Americans, and they worked at their improvements across a broad front of craft and trade enterprise. The other observation to be made is that while most African-Americans in the late nineteenth century still lived in rural areas and worked in agriculture, those with a particular interest and talent for invention gravitated to urban and industrial centers where there were more jobs and better pay. And it was in those places that the most well-known black inventors sharpened their gifts.

Jan Matzeliger (1852–1889) is a good illustration of the case. He apprenticed as a machinist in Surinam, his native country, and then immigrated to New England. Unable to find work as a machinist, he took a series of odd jobs, including one sewing on shoe soles in a factory in Lynn, Massachusetts, then the shoe-making center of America. Working in that industry led him to a mechanical solution for a particular problem in making shoes. Machines that could sew and tack shoes already existed, but the painstaking job of pulling, smoothing, and shaping leather until it approached the form of a human foot still had to be done by hand. That technique, called lasting, represented a bastion of craft skill from the worker's point of view, but from the manufacturer's perspective a considerable bottleneck to fully mechanized shoe production.

Matzeliger worked on several models of a lasting machine until he had produced a version that so approximated his final goal that he decided to patent it. However, this working model had taken four years to develop, he was suffering physically as well as financially, and he needed still more money to perfect the model, test it, and apply for the patent. In the end, he was forced to sell two-thirds of his ownership in the patent rights to local investors—not an uncommon plight for poor inventors, especially as it cost ever more to produce improvements in increasingly complex technologies.

In its trial run Matzeliger's model successfully performed the lasting of seventy-five pairs of shoes, but he continued to refine his ideas and produced two more machines that were improvements on the original. Ironically for blacks, so often themselves the victim of technological unemployment, disgruntled shoe workers dubbed Matzeliger's invention the "Niggerhead" machine.[31] And indeed, after its usage became widespread, the craft of shoe lasting practically disappeared. An extraordinary technical accomplishment in itself, this invention essentially completed the mechanization of the shoe industry. The United Shoe Machinery Company, striving for monopoly control of the machines that made shoes, eventually bought Matzeliger's patents, but he died in poverty.

The most significant mechanical industry for the employment of African-American males in the post Civil War period was the railroad. During the 1880s more than 70,000 miles of railroad track were laid, making rail transportation the country's largest industry. Black men worked not only in the more visible service positions, but also in the most dangerous and arduous jobs. This rapid expansion of the railroad system called into existence many technological advances, and the Patent Office issued a wide array of patents to black inventors.

Humphrey H. Reynolds, for example, patented a ventilator in 1883 for passenger cars that allowed in fresh air, but kept out the dust and soot that usually enveloped a moving car. The *Baltimore Afro-American* reported on Reynolds's ventilator when it was displayed at the Atlanta Exposition:

The H. H. Reynolds ventilator in the Pullman cars is perhaps the most widely used of those exhibited at Atlanta. Reynolds was a porter on one of the Pullman cars. Opening and shutting the windows as he did so often for his passengers, he devised a screen to keep the cinders out. Pullman heard of it and Reynolds was sent for. He explained his invention to the car magnate, and the interview resulted shortly after ward in the adoption of this ventilator on all the Pullman cars. Reynolds claimed the invention, but Pullman did not recognize the claim. He got out of the service of the Pullmans, sued them, and got a verdict for ten thousand dollars.[32]

The story of Andrew Beard, another black inventor, also had a happy ending. He worked in a railroad yard in Eastlake, Alabama, and had occasion to notice the difficulties in manually coupling cars together. It was a dangerous job, the cause of many injuries, and also the source of jarring lurches and jolts when trains started or stopped. Indeed, this common problem became one of the most popular subjects for inventors. In 1897 alone 6,500 different kinds of couplers were invented, and by 1930 11,813 patents for them had been issued. One of those in 1897 was Andrew Beard's, and he sold the rights to it for $50,000.[33] But not all inventors fared so well. It took money to fight off challenges to a patent, and few black inventors had the resources for a protracted legal battle.

Railroad locomotives were another particular focus of inventive activity, and they attracted the attention of Elijah McCoy (1843–1929), one of the most prolific of nineteenth-century African-Americans inventors. He was born in Colchester, Ontario, in a community of escaped slaves, and went to Scotland as a young man where served a mechanical engineering apprenticeship in Edinburgh. McCoy then migrated to the United States, and wound up seeking a position in Ypsilanti, Michigan, the headquarters of the Michigan Central Railroad. Despite his training, the only job available to him was as a locomotive fireman but it soon made him aware of the problems of overheating common to steam locomotives, since so many moving parts could not be lubricated while the engine was in motion. In 1872, McCoy patented the first of his automatic lubrication devices—one that turned out to be widely employed on stationary steam engines that were used in factories. He assigned his rights to this patent to two Ypsilanti investors, using the money he received for further studies of lubrication problems. The Michigan Central also gave him a new job as an instructor in the use of the new lubricators, which were widely adopted by railroad and shipping lines. Over the next several years, McCoy patented more than fifty additional inventions, most of them related to lubrication. He later moved to Detroit, where the Elijah McCoy Manufacturing Company had been established by the assignees of his most valuable patents.[34] McCoy, however, was not a major stockholder and later in life suffered a number of financial setbacks. After creative insight, the next best knowledge for an inventor to possess is an understanding of the economics of patent management.

That post-emancipation outpouring of inventiveness, and the dramatic surge of American industrial development in the 1870s and the 1880s, persuaded many African-American leaders that technical and industrial training was the key to their full incorporation into the nation's life. The success of individual black inventors was thus held up as a beacon of what might be achieved on a wider scale. Expositions and fairs had long been a popular means of demonstrating the achievements of blacks to the wider public. But after the Civil War, they took on the specific purpose of exhibiting the "phenomenal progress of the colored American" since emancipation. Several states sponsored "emancipation expositions" on the anniversaries of the Emancipation

Proclamation. African-Americans continued to organize independent expositions and fairs; they also participated in the state and regional expositions where black crafts and manufactures, as well as inventions by blacks, were prominently featured in segregated "colored" or "Negro" departments. Thus, inventions by blacks were displayed at the Cotton Centennial in New Orleans in 1884, the Chicago World's Fair in 1893, and the Southern Exposition in Atlanta in 1895. Benjamin Montgomery, for example, now a free man, proudly exhibited at the Western Sanitary Fair in Cincinnati the propeller he had invented while a slave of Joseph Davis—the brother of Jefferson Davis, President of the Confederacy.[35]

In an 1894 speech before the House of Representatives on behalf of proposed legislation sponsoring another Cotton States Exhibition to publicize the South's economic and technological progress since the Civil War, Representative George Washington Murray (1853–1926) read into the *Congressional Record* the names and inventions of ninety-two African-American inventors. Murray himself was responsible for twelve of those inventions, but the point of his remarks was to urge that a separate space be reserved to display some of the achievements of Southern black people, and he set out the reason why they wanted to participate in such expositions:

Mr. Speaker, the colored people of this country want an opportunity to show that the progress, that the civilization which is now admired the world over, that the civilization which is now leading the world, that the civilization which all nations of the world look up to and imitate— the colored people, I say, want an opportunity to show that they, too, are part and parcel of that great civilization.[36]

Murray's list was derived from the research of Henry E. Baker, a Patent Office examiner who dedicated most of his life to uncovering and publicizing the contributions of black inventors. Baker's research also provided the information used to select black inventions exhibited at the Cotton Centennial in New Orleans, the Columbian Exposition in Chicago, and the Southern Expositions in Atlanta. By the time of his death, Baker had compiled four massive volumes of patent drawings and specifications for patents awarded to black inventors.

In 1900, at the request of the U.S. Commission to the Paris Exposition, and with Baker's assistance, the U.S. Patent Office sent letters to more than 3,000 patent attorneys, manufacturers and newspaper editors asking them to list any black inventors who might have come to their attention. This was the government's first systematic effort to collect information about inventions by blacks. The results revealed the names of more than 400 inventors who had received patents, and of many more who had tried to obtain patents. This data found its way into the U.S. exhibit's Negro Department, organized by Thomas J. Calloway and W. E. B. Du Bois which, among other materials about technical and industrial training, displayed 350 new patents granted to black inventors.[37] And the Jamestown Exposition, held in Jamestown, Virginia, in 1907 to commemorate the tercentenary of European settlement there, provided yet another opportunity to call attention to the technological progress of the race. A separate building designed by black architects and constructed by black contractors housed displays of the work of black artisans, inventors, and students.[38]

While always cast in positive language, these multiple efforts to promote and celebrate African-American inventiveness were also designed to attack the myth of black intellectual inferiority, so explicitly advanced in those years. Newspapers like the *Southern Workman, Colored American Magazine,* and *Crisis* trumpeted the contributions of black inventors, and so did a variety of other publications such as Munroe Work's *The Negro Yearbook, Twentieth-Century Literature,* which D. W. Culp edited, and *Evidences of Progress among Colored People* by G. F. Richings. The search for recognition was important, Henry Baker explained in 1902:

Judging from what has been duly authenticated as Negro inventions patented by the United States, it is entirely reasonable to assume that many hundreds of valuable inventions have been patented by Negro inventors for which the race will never receive due credit. This is the more unfortunate since the race now, perhaps, more than ever before, needs the help of every fact in its favor to offset as far as possible the many discreditable things that the daily papers are all to eager to publish against it.[39]

By 1900 more black people were living in cities than ever before. Thousands had left the fields and wash-tubs of the South to seek their fortunes in the factories and laundries of the North. Madame C. J. Walker (1867–1919) was one of those migrants, and she prospered from innovative cosmetic products marketed to this new urban black population. Born Sarah Breedlove in rural Louisiana, she settled in St. Louis, Missouri, supporting herself and her daughter by working as a washer-woman. But she had an idea for beauty products, and by 1905 she had created a formula for straightening and grooming black women's hair. She first sold these hair preparations in Denver, Colorado, under her married name, but after five years of aggressive salesmanship she was able to establish a headquarters in Indianapolis for the national distribution of her goods.[40] Two strategies proved especially successful; she developed a whole system of hair and cosmetic products, and then established salons of her own that popularized straightened hair as the image of a smart, effective city woman.

Just as Harlem became a Mecca for ambitious African-Americans in the early twentieth century, and a center of black culture and personal style, Washington, D.C., was a popular destination for upwardly mobile blacks because of the employment opportunities that the federal government offered. Several Washington residents who worked as civil servants patented successful inventions. Robert Pelham (1859–1943) began his career as a newspaper publisher and editor in Detroit. His paper, the *Plaindealer*, became one of the most successful black newspapers of the Midwest, and that led him into Republican party activities. Perhaps because of that connection, he moved to Washington in 1900 and began a thirty-seven-year tenure at the U.S. Census Bureau. During the course of his work he conceived and patented a tabulating machine in 1905, and an adding machine device in 1913.[41] Another Washingtonian, Shelby Davidson (1868–1931), came from Lexington, Kentucky, in 1887 to work in the auditing department of the U.S. Post Office Department, and he was also drawn into the technology of office equipment. In 1906 Davidson began to study adding machines with an eye to improving them to handle government auditing functions more efficiently. He visited several factories to observe exactly how they constructed these machines, and after two years of study patented his first invention, a rewind device for calculators. Davidson claimed that "by the use of this device the government would save three-fourths of the paper used on the machines and the time of the clerks in taking up the paper."[42] Fascinated by the problem of mechanical tabulation, Davidson worked on various improvements for his device, and in 1911 received another patent, this one for an "automatic fee device" that helped postal clerks assess the correct fees.

Besides the difficulties and expenses of the creative process itself—working out the design issues, contriving models to demonstrate the principles involved, preparing written descriptions that provide the appropriate legal basis for the claim advanced, and getting made the detailed drawings that illustrate all the properties of the invention—inventors had also to think about manufacturing and marketing their creations, and about securing their rights against encroachment. And the more valuable the patent, the more effort and money were required to protect it. By the 1900s manufacturing and marketing a new invention presented such formidable problems that many inventors chose to assign their patents outright to agencies, private investors, or corporations. That gave inventors an immediate financial reward for their inventions and freed them from the burden of risk, but they lost all rights to whatever profits the patent subsequently earned for its new owners. More often than not, money concerns decided the issue for poor inventors, and they were the ones most likely to assign their patents.

But Samuel Scottron, a New York manufacturer, provides the example of someone determined to market his own inventions. Originally a merchant of household goods, Scottron first designed an adjustable mirror. These mirrors, he claimed, were "so arranged opposite each other as to give the view of every side at once . . . and so simple too, that it was impossible to accomplish the same result in a simpler or cheaper manner."[43] Unwilling to have someone else to make his mirrors, or to assign his patent,

Scottron decided to manufacture them himself. To do that, he enrolled in night school where he studied practical mathematics, also apprenticing himself to a pattern-maker and to a master mechanic, and then began to make his mirrors. They proved successful, and he then patented other versions that he also produced.

Having learned to manipulate materials, Scottron left his mirror-making business behind and became interested in the manufacture of window cornices. He described the rise and decline of this enterprise for the readers of the *Colored American Magazine*:

> Very soon, however, a new patent having great possibilities was granted me, for an extension cornice. I received several patents for these and abandoned the mirror, putting these out on a royalty, and entering the manufacture of extension cornices, which coined thousands while it lasted, an excellent thing in every way; but it came to grief through one of those causes that will sometimes lay out, stiff dead, the best thing in the market, viz.: the capriciousness of fashion. Curtain poles came into fashion and killed the cornice business entirely, in less than six months of activity in opposition.[44]

Undaunted, he abandoned the extension cornice business to begin manufacturing synthetic onyx, an endeavor in which he enjoyed considerable financial success until he retired.

Scottron was a tireless booster of black business and invention, wrote several articles encouraging people to go into manufacturing and trade, and offered advice to would-be inventors as well. First he suggested that patent holders try to manufacture their own inventions, and to learn as much as possible about design and mechanical engineering:

> There is possibly no shop where one can serve and get a broader knowledge of applied mechanics than a well patronized pattern-making shop, bringing one as it does into a consideration of the various elements, substances; etc. used in manufacture; their nature and possibilities. . . . It grounded me and gave me confidence in myself, and an actual knowledge of possibilities, which prevented many costly ventures and foolish mistakes, such as the patenting of things absolutely useless.[45]

Scottron also recommended that the prospective inventor aim for simplicity in design and construction. Simplicity, he claimed, is "a thing very necessary in patent articles. A patent which can be simplified by another is worth nothing." The key to simplicity was mechanical knowledge, understanding "how not to use three motions where two will do the work." The other thing to know about was the market. Familiarity with tastes and demand "will show you whether what you wish to accomplish will be worth anything in the market."[46]

For those attempting a career as a full time inventor, getting into production and marketing could seem a next logical step. Garrett Morgan (1875–1963) invented and patented a safety hood in 1914, and gained national prominence for his device when he used it to rescue workers trapped by fire and gases in a tunnel explosion at

Cleveland, Ohio. It was quite a dramatic moment. Two previous rescue efforts by police and firemen ended in disaster when additional gas explosions killed nine of them. Morgan then arrived on the scene with several of his safety hoods, which he took into the tunnel and used to bring out three survivors, as well as the bodies of the others.[47] The resulting publicity brought investors, and together with Morgan they formed the National Safety Device Company to manufacture his hoods. In 1923 Morgan patented another safety invention, a mechanical traffic signal, which came into wide usage. He marketed the signal through a company he formed for that purpose, the G. A. Morgan Safety System, but then later assigned the patent to the General Electric Company.

The years from 1880 to 1930 seem the heroic age of American invention because so much technical advance—particularly in the fields of electricity and communication—can be associated with individuals. The names that come most obviously to mind are those of Edison and Bell, with Westinghouse, Sprague, and Maxim not far behind in celebrity. We also know them because they were highly successful entrepreneurs, and their names survive in the corporations they established. Less celebrated by the popular press, but certainly worth knowing about are African-American inventors like Granville Woods and Lewis Latimer, both also associated prominently with the electrical industry.

Granville Woods (1856–1910) was born in Columbus, Ohio, where he apprenticed as a machinist and blacksmith. He then worked for a time as a railroad fireman and engineer, but became quite interested in the emerging field of electrical engineering, began reading deeply in the subject, and took evening courses in electrical and mechanical engineering. The need for a job led him back to steam engines, working on a British steamer for a couple of years, and again on American railroads—employment that also resulted in his first patent, in 1884, for an improved steam boiler. But, with the model of Edison clearly in mind, he seemed determined to devote himself to electrical inventions, and founded the Woods Electric Company in Cincinnati to research, manufacture, and market his inventions. The first of his electrical inventions, for an improved telephone transmitter, came shortly afterward and was patented in 1884.[48]

In fact, Woods did not actually have the capital to develop his first two inventions, and he assigned his patent rights to others. Furthermore, in the case of his telephone transmitter he faced stiff competition from Alexander Graham Bell, who already had a well established company to produce telephone equipment and the protection of prior patents. Still, the improvement of known devices is a staple feature of invention, and in 1885 Woods patented another "electrical apparatus for transmitting messages." He sold this device, which allowed operators to transmit either Morse code or voice messages, to the Bell Telephone Company.

Woods continued to design and patent electrical equipment, particularly for railway telegraph and electrical railway systems. These inventions brought him into direct competition with Thomas Edison and another inventor named Lucius Phelps, who had invented similar telegraphic devices, and which of them would enjoy patent protection for the invention had to be determined in court. Woods won, and his legal victory also brought publicity for his career as an inventor, which the *American Catholic Tribune* described:

Mr Woods, who is the greatest electrician in the world, still continues to add to his long list of electrical inventions. The latest device he invented is the synchronous multiplex railway telegraph. By means of this system, the railway dispatcher can note the position of any train on the route at a glance. The system also provides for telegraphing to and from the train while in motion. The same lines may also be used for local messages without interference with the regular train signals. The system may also be used for other purposes. In fact, 200 operators may use a single wire at the same time. Although the messages may be passing in opposite directions, they will not conflict with each other. In using the device there is no possibility of collisions between trains as each train can always be informed of the position of the other while in motion. Mr. Woods has all the patent office drawings for these devices as your correspondent witnessed.

The Patent Office has twice declared Mr Woods prior inventor. The Edison and Phelps Companies are now negotiating a consolidation with the Woods Railway Telegraph Company.[49]

Woods moved to New York City in 1890, to take advantage of the better opportunities for electrical engineers there, and went on to develop a number of improvements in the equipment used in electric street car systems. Among those innovations were the application of secondary "dynamotors" that reduced the risk of fires, and a technique for better connecting the street car to its electric power system. That device, called a troller, consisted of a grooved wheel at the end of a wand under spring tension, that pushed against the overhead electrical wire and so with less friction loss conducted current to the motor of the trolley car.

In an earlier era, when most inventions grew simply out of familiarity with craft practice and work experience, the chief obstacle for black inventors, besides racial prejudice, was money. But by the twentieth century, both technology and its institutional structures had changed. Many of the most important innovations now began to come from teams of researchers employed by large corporations that assumed the rights to their patents. More and more often, inventors were salaried personnel and had university degrees in science or engineering. Few black men and women were able to obtain the necessary degrees in those fields of study, and those who did had to overcome the reluctance of most firms to employ them in such positions.

But a small number, due to their talent and persistence, did manage to find places in large-scale industrial research and development enterprises. Lewis H. Latimer (1848–1928) was one of them. Born in Boston, he was the son of an escaped slave,

George Latimer, who became famous for the defense campaign mounted on his behalf by the abolitionist William Lloyd Garrison. When George Latimer's owner appeared in Boston, demanding the return of his "property," the abolitionists of the city staged a series of rallies and fund-raising events to raise enough money to purchase Latimer's freedom.

Lewis Latimer served the Union cause in the Civil War. After his discharge, he took a job as office boy in a Boston patent law firm. But his employers were so impressed with his drawing abilities that he soon became a patent draftsman, and then the head draftsman for the firm. Latimer made the patent drawings for many of Alexander Graham Bell's telephones, and testified in court on Bell's behalf when his patents were challenged. Next he became a patent draftsman at Hiram Maxim's United States Electric Company and began more closely to work with the developing electrical technology. Maxim was one of Edison's biggest competitors, and much of the struggle for the burgeoning market for electrical lighting revolved around the search for improved light bulb filaments. In 1881, one year after joining Maxim's firm, Latimer and a co-worker patented an improved method for bonding carbon filaments.

The following year, Latimer patented a technique for making carbon filaments. This was one of his most significant inventions, for carbon filaments produced by his method were much more cost efficient.[50] As Latimer's expertise increased, he was given more responsibilities, and was soon supervising the installation of electric light plants in New York, Philadelphia, and other cities. He also traveled to London to establish a department in Maxim's branch there for the production of his light bulb filaments.

In 1884, Latimer began working with the engineering department of the Edison Electric Light Company in New York. Six years later, he was transferred to Edison's legal department, where he served as chief draftsman for patents under legal dispute. His skill in illustrating electrical patents led 1896 to his appointment as draftsman for the Board of Patent Control, a body established by General Electric and Westinghouse to oversee patent disputes between them. Latimer worked with that group until 1911, when he left to practice as a patent consultant in New York. In an evocative self portrait (figure 1), Lewis depicted the difficulty, at his age, of deciding between job security with corporate enterprise, or the economic risks of independent consulting.

These sketches of African-American inventiveness, brief as they are, reveal some important truths. One is that black people, both women and men, have been active participants in the history of American technology from the very beginning. Even enslaved, they were moved to create improved ways of doing things. Indeed, inventiveness was agency, a means of taking as much control of one's destiny as possible, and there is

Figure 1
Lewis H. Latimer, "My Situation as It Looked to Me in 1912." Courtesy of Latimer-Norman Family Collection, Queens Borough Public Library. In this blueprint drawing, Latimer depicts himself struggling to choose between the freedom of a career as a technical consultant and the job security of corporate employment. But the drawing might also be taken to represent the difficulties blacks faced in finding jobs as engineers.

plenty of evidence that a great many grabbed at the chance. Their ideas, as well as their labor, also proved a source of great wealth to eighteenth- and nineteenth-century America, and that is worth remembering, too.

But perhaps the most important lesson of all is that their stories give the lie to all those old notions of inferiority. Denied the advantages of formal education or university degrees, without the funds to amplify inspiration, and against a strong tide of ingrained ill-will, these African-Americans proved capable of sustained and creative technical accomplishment. As imperfectly as we know their history, that much is certainly true.

Notes

1. Theresa Singleton, "Buried Treasure," *American Visions*, March-April 1986, p. 36.

2. "Old Pipes Offer New View of Black Life in Colonies," *New York Times*, July 12, 1988.

3. Ibid.

4. John Vlach, *The Afro-American Tradition in the Decorative Arts* (Cleveland Museum of Art, 1978), p. 102.

5. Peter Wood, *Black Majority: Negroes in Colonial South Carolina from 1670 through the Stono Rebellion* (Norton, 1975), p. 25. See also Dale Rosengarten, *Row upon Row: Sea Grass Baskets of the South Carolina Low Country* (McKissick Museum, 1986), p. 6.

6. O. J. E. Stuart to Secretary Jacob Thompson, August 25, 1857, Office of the Secretary of the Interior, Records Concerning the Patent Office, 1849–89, Letters Concerning the Patent Office, vol. 1 (1857–73), National Archives.

7. Ibid.

8. Stuart to Thompson, December 18, 1857.

9. Gary Nash, "Forging Freedom: The Emancipation Experience in the Northern Seaport Cities, 1775–1820," in *Slavery and Freedom in the Age of the American Revolution*, ed. I. Berlin and R. Hoffman (University Press of Virginia for U.S. Capitol Historical Society, 1983), p. 8. More specifically, see W. Jeffrey Bolster, *Black Jacks: African American Seamen in the Age of Sail* (Harvard University Press, 1997).

10. James B. Farr, Black Odyssey: The Seafaring Traditions of Afro-Americans, Ph.D. dissertation, University of California, Santa Barbara, 1982, p. 31.

11. Portia P. James, *The Real McCoy: African American Invention and Innovation, 1619–1930* (Anacostia Museum and Smithsonian Institution Press, 1989), pp. 33–35.

12. Sidney Kaplan, "Lewis Temple and the Hunting of the Whale," *New England Quarterly* 26 (1953), March, p. 85.

13. William J. Simmons, *Men of Mark: Eminent, Progressive, and Rising* (Arno, 1968 reprint of 1887 edition), p. 1012.

14. Winthrop Jordan, *White Over Black: American Attitudes Toward the Negro, 1550–1812* (Penguin, 1968), p. 406. Speaking of the deep animosity of white craftsmen towards black workers, Frederick Douglass complained that it would be easier to get his son "into a lawyer's office to study law than . . . into a blacksmith's shop to blow the bellows and to wield the

sledgehammer" (Wilson J. Moses, *The Golden Age of Black Nationalism, 1850–1925*, Oxford University Press, 1978, p. 28).

15. See Jennings's obituary in *Anglo-African Magazine*, April 1859 (Arno and New York Times, 1968 reprint).

16. Elizabeth Keckley, *Behind the Scenes; Thirty Years as a Slave and Four Years in the White House* (Arno and New York Times, 1968 reprint).

17. William C. Nell, *Colored Patriots of the American Revolution* (Arno, 1968 reprint of 1855 edition), p. 265. See also Jane Sikes, *The Furniture Makers of Cincinnati, 1790 to 1849* (Cincinnati, n.d.).

18. Rayford W. Logan and Michael R. Winston, eds., *Dictionary of American Negro Biography* (Norton, 1982), p. 481.

19. Thomas Jefferson, *Notes on the State of Virginia* (University of North Carolina Press, 1955), pp. 139–142.

20. As quoted in Silvio Bedini, *The Life of Benjamin Banneker* (Scribner, 1972), p. 152.

21. Charles Wesley and Abram Harris, *Negro Labor in the United States, 1850–1925* (Russell & Russell, 1967 reprint of 1927 edition), p. 51.

22. Logan and Winston, *Dictionary of American Negro Biography*, p. 485.

23. Louis Haber, *Black Pioneers of Science and Invention* (Harcourt Brace and World, 1970), pp. 13–23. See also Patricia and Frederick McKissack, *African American Inventors* (Millbrook, 1994), pp. 41–48. The Louisiana State University Library maintains a very useful website called "Faces of Science: African Americans in the Sciences," and it has technical details on Rillieux's inventions.

24. "News Notes: Items of Interest Concerning Our People Everywhere," *Baltimore Afro-American*, October 26, 1895.

25. McKissack, *African American Inventors*. Henry Baker, *The Colored Inventor: A Record of Fifty Years* (Arno, 1968 reprint of 1915 edition) also contains basic information about a wide array of black inventors.

26. The Johns Hopkins University Applied Physics Laboratory maintains a web site with information on African-American inventors and scientists, including Hyde.

27. For more information about women inventors, see Autumn Stanley, "From Africa to America: Black Women Inventors," in *The Technological Woman: Interfacing with Tomorrow*, ed. J. Zimmerman (Praeger, 1983).

28. Michigan Freedmen's Progress Commission, *Michigan Manual of Freedmen's Progress* (J. M. Green, 1968 reprint), pp. 41–44.

29. Henry Baker, *The Colored Inventor* (Arno and New York Times, 1969 reprint), p. 9. See also C. R. Gibbs, *Black Inventors from Africa to America* (Three Dimensions, 1995), p. 96.

30. Baker, *The Colored Inventor*, p. 9.

31. Sidney Kaplan, "Jan Earnest Matzeliger and the Making of the Shoe," *Journal of Negro History* 40 (1955), January, p. 21.

32. "Some Afro-American Inventors: Patents That Have Been Taken Out by Colored Men," *Baltimore Afro-American*, November 2, 1895.

33. McKinley Burt Jr., *Black Inventors of America* (National Book Company, 1989), pp. 1, 18–20. See also Patricia Carter-Ives, *Creativity and Inventions: The Genius of Afro-Americans and Women in the United States and their Patents* (Research Unlimited, 1987), pp. 29–31.

34. James, *The Real McCoy*, pp. 73–83.

35. Moses, The Gold*en Age of Black Nationalism*, p. 46. Not everyone agreed on the subject of black participation in segregated exhibits, however. See, for example, Ida B. Wells-Barnett, *The Reason Why the Colored American Is Not in the World's Columbian Exposition* (Chicago, 1893); Linda McMurry, *The Life of Ida B. Wells* (Oxford University Press, 1998), p. 186; and Robert Rydell's version of the events in his *The Reason Why the Colored American Is Not in the World's Columbian Exposition: The Afro-American Contribution to Columbian Literature* (University of Illinois Press, 1999).

36. *Congressional Record*, 53rd Congress, 2nd session, August 10, 1894, p. 8382.

37. W. E. Burghardt Du Bois, "The American Negro at Paris," *American Monthly Review of Reviews*, pp. 22, 575.

38. Lucy Brown Franklin, "The Negro Exhibition of the Jamestown Ter-Centennial Exposition of 1907," *Negro Historical Bulletin* 38 (1975), June-July, p. 408. See also Ruth Winton, "Negro Participation in Southern Expositions, 1881–1915," *Journal of Negro Education*, winter 1947, p. 34; Giles Jackson and D. Webster Davis, *Industrial History of the Negro Race* (Books for Libraries Press Reprints, 1971 reprint of 1911 edition); Francis H. Warren, *Michigan Manual of Freedmen's Progress* (Freedmen's Progress Commission, 1915).

39. Henry Baker, "The Negro as an Inventor," in *Twentieth-Century Negro Literature*, ed. D. Culp (Arno, 1969 reprint of 1902 edition), p. 401.

40. James, *The Real McCoy*, p. 85.

41. Michigan Freedmen's Progress Commission, *Michigan Manual of Freedmen's Progress*, pp. 87–92. See also Baker, *Colored Inventor*, p. 10.

42. Autobiographical statement in the Shelby Davidson Papers, Moorland-Springarn Research Center, Howard University.

43. Samuel Scottron, "Manufacturing Household Articles," *Colored American Magazine*, October 1904, p. 662.

44. Ibid., p. 623.

45. Ibid., p. 622.

46. Ibid., p. 623.

47. Morgan provided a compelling description of this rescue in a letter to F. M. Wilmot. Garrett Morgan Papers, Western Reserve Historical Society, Cleveland.

48. Aaron Klein, *Hidden Contributors: Black Scientists and Inventors in America* (Doubleday, 1971), p. 73.

49. As quoted in Klein, *Hidden Contributors*, p. 73.

50. James, *The Real McCoy*, pp. 97–99. See also Klein, *Hidden Contributors*, p. 71.

History in the Funny Papers

Bruce Sinclair

We use humor to talk about serious things, so we know it can be an interesting tool of analysis. When black comic strip artists became nationally syndicated, they naturally took advantage of their opportunities to address themes important to African-Americans. One ubiquitous subject in all of these strips is invention and the contribution of black inventors. And the plainly didactic character of some of the examples shown here tells us that their artists were concerned to educate both blacks and whites.

The seven examples of cartoon art reprinted here clearly reveal the cultural value of invention to African-Americans, but they also suggest the problematic nature of representing black ingenuity in a society that so long denied its existence.

Morrie Turner, of Oakland, California, a self-trained artist, is said to have integrated the comics with his "Wee Pals," which emerged along with the Civil Rights movement of the 1960s, found increasing numbers of readers after Martin Luther King's assassination, and became nationally syndicated three years later. Ray Billingsley, the creator of "Curtis," came from Wake Forest, North Carolina, to Harlem, and studied at Manhattan's School of Visual Arts. He bases his cartoon character's experiences on his own memories of growing up in a big city, where besides having to negotiate the terms of co-existence with their parents, kids have constantly to deal with the complexities of their neighborhood. Robb Armstrong, who draws "JumpStart," was born in West Philadelphia, one of five children raised by a single mother, and he graduated from Syracuse University with a degree in fine arts. Like Turner and Billingsley, Armstrong unambiguously presents black inventors as a crucial index of African-American accomplishment, in a context that emphasizes close family relations and multi-cultural understanding.

Aaron McGruder's "Boondocks" gives us a more complex presentation of the role of invention, in a comic strip that is itself complicated by the situation of black kids from the urban ghetto whose families have moved them to the white suburbs. A University of Maryland graduate in Afro-American Studies, McGruder says of himself: "He has no BMW, no Benz, not a single article of clothing from Versace or FUBU, and no life as a result." Sharp-edged and often sardonic, the strip still uses invention as a way of talking about black-white relations, and still depends for its humor upon simplistic notions about African-American inventiveness.

This highlighting of invention in our comic strips is typically American; in the history of cartoons inventing and inventors is one of the big staples of subject matter—from Betty Boop (Grampy was an inventor) to *A Bug's Life* (Flick hopes to invent a solution to the colony's problems). African-Americans are equal sharers in this national fascination, of course, but because they have been left out of that part of our history they have in their own comic strips a double reason for claiming technological competence.

Acknowledgments

"Wee Pals" is reprinted with permission of Morrie Turner and Creators Syndicate, Inc. "Curtis" is reprinted with special permission of King Features Syndicate. "Jumpstart" is reprinted with permission of United Features Syndicate, Inc. "Boondocks" (©2000 Aaron McGruder) is reprinted with permission of Universal Press Syndicate (all rights reserved).

WEE PALS/ by Morrie Turner

CURTIS / by Ray Billingsley

CRISPUS ATTUCKS (1723-1770)-"FIRST TO DIE FOR INDE-PENDENCE"... GRANVILLE T. WOODS (1856-1910)"PROLIFIC INVENTOR"... DANIEL HALE WILLIAMS (1858-1931)"FIRST SUCCESSFUL HEART SURGEON" --I'VE NEVER HEARD OF ANY OF'EM!--AND THEY WERE BLACK!!

SIMMER DON', MON! CAN'TCHA SEE I'M TALKIN'T' ME HONEYPOT?.--HERE. READ.

UM... MY WIFE WOULD LIKE SOME WATER!

AH, MUS'TAVE BEEN DE TIDES OF GOOD FORTUNE TO BRING YOU HEAH TO MISS RUBY'S DOOR, HONEYPOT! MISS RUBY, SHE NOT EVER LEED YA WRON'!

I CUT OUT A BUNCH OF CLIPPINGS FROM PEOPLE, EBONY, AND ENTERTAINMENT... FIGURE I'D WRITE ON TINA TURNER-Y'KNOW, MAYBE TONI BRAXTON.

CAN WE GET SOME WATER OVER HERE?

SAY, WHAT'TIS ALL DIS YOU SO NONCHALANTLY HAVE SPLAYED ALL ON ME KLEEN COUNTER, HONEYPOT? TELL MISS RUBY

MY STUFF FOR BLACK HISTORY MONTH, MISS RUBY

FIRST TIME A WAITRESS EVER GAVE ME A TIP!

OH, LET ME ATTEND DIS MAN'FORE HE RAISE DE ROOF OFF DE BUILDING!

HEY, WATER! WATER OVER HERE FOR MY WIFE!!

YOU BORRY MISS RUBY'S BOOK, WRITE YO' REPORT, AND WHEN YOU RECEIVE A MASTER'S GRADE, YOU TODDLE ON BACK TO SEE MISS RUBY!

THANKS, MISS RUBY!!

ON JAMAICA, DERE IS DIS SAYING, "A FISHERMAN KNOWS DE SHALLOWS CONTAINS MANY FISH, BUT DE SMART FISHERMAN KNOWS T'COTCH WHALES, ONE 'AS TO GO DEEP'".

THE WAY THE MEDIA HAS US BRAINWASHED, THEY HAVE US THINKING SUCCESS IS MEASURED BY ATHLETIC ABILITY OR ENTERTAINMENT SKILLS.

JUMPSTART

JUMPSTART

THE BOONDOCKS *Aaron McGruder*

New South, New North: Region, Ideology, and Access in Industrial Education

Nina Lerman

The young man's story was exactly the kind Fanny Jackson Coppin hoped to be able to tell when she began fighting in the early 1880s to add an Industrial Department to the renowned academic programs at the Institute for Colored Youth (ICY) in Philadelphia. Working steadily shucking oysters, he had earned $7 a week. He would gladly have made a beginning in, for example, the construction trades, learning techniques from his elders on the job, as a young white man might do—but such a path was closed to African-American youth in 1891. Instead he completed the new ICY course in bricklaying, and demonstrating the success of Coppin's strategy, he found he could make $3.25 per day, potentially as much as $19.50 in a standard six-day week. Coppin, a community activist as well as the scholarly principal of ICY, had recognized and solved—in a small way, but one she hoped would expand—the complex economic problem of access to technological knowledge.[1]

Coppin did not conflate the purposes of the new Industrial Department with the mission of her high school teaching faculty: ICY was a well-established academic high school, boasting a highly educated cadre of African-American teachers and staff dedicated to training young black scholars—the kinds of students who went on to attend college, to sail past white competitors on the city's teachers' exams, and to "improve the race" in a range of professions.[2] A graduate of Oberlin College herself, she expected trades training to serve a different group of students, and her justifications for it were economic, not intellectual. Coppin spent several years convincing the white board of managers to fund this new undertaking, a departure for both the institution and the city. Nonetheless she and the new industrial teachers, who included the first white teacher hired at ICY, succeeded by the end of the 1880s in creating a facility so popular that its programs drew a chronic waiting list.[3] As in its academic curricula, Coppin's ICY redirected the paths of opportunity: Where else could a black child learn Latin and physics? Who else would teach an ambitious young oyster shucker how to mix mortar and lay bricks, or a housemaid the intricacies of dressmaking?

But in the United States these were decades of change in educational philosophy, and of hardening in racial ideology. By 1902, when Coppin retired, the prevailing winds had shifted so completely that the managers, far from resisting trades training, *entirely* recharted the ICY's mission: they closed down the high school in favor of an "industrial school" on a newly purchased farm in the country, where the academic program was limited in favor of "domestic science," woodworking, and metalworking. "Industrial education," for both men and women, would train teachers for the "colored schools" of the rural South. The African-American scholars of the ICY faculty were left to find work elsewhere and a new faculty hired. The school's new principal, recommended by Booker T. Washington of Tuskegee Institute, had not long before served on the faculty of Washington's alma mater, the famous Hampton Normal and Agricultural Institute in Virginia.

Making Education "Industrial"

Historians of education and of the African-American experience will recognize familiar contours in this story, and know well the severe limitations on academic programs standard in schools following the "Industrial Education" model provided by Hampton. The Institute's emphasis on hard work, Christian morals, and self-help without systemic social change produced a program so circumscribed one historian has called it "schooling for the new slavery." Similarly, the long reach of Booker T. Washington's influence after the mid 1890s—often dubbed the "Tuskegee Machine"—comes as no surprise.[4] But the ICY Industrial Department and the relocated Industrial School illustrate only two of the possible meanings of "industrial" in this period. The slippery reappearance of this often anomalous term in such tales has gone largely untold. What made industrial education *industrial*? When did it refer to a character-building industriousness (industrial as opposed to idle, a common meaning in the mid-nineteenth century), and when to an economically valuable training (industrial as opposed to rural, primitive, backward)?[5] And when to milking cows, and when to running electric plants? Industrial Education, as James Anderson has shown, most often meant a drastically reduced literary education.[6] But to leave the analysis there misses a substantial part of the story.

The technological content of the various trades and manual training programs must be carefully studied in the context of contemporary ideas about education, about technology, about work, about race. Further, the apparent transfer of technique from South to North needs to be examined in light of our understandings of regionalism in this period: the expected north-to-south direction of industrial influence; the chronic relegation of African-American workers in Northern cities to service work rather than to industrial production jobs; discussions of labor and education in the New South. We have barely begun exploring the various meanings—both cultural and curricular—of words like "industrial" and "trades" in racialized educational contexts, or in regional perspective.

Nationally, professional educators' ideas about the relationship between schooling and "daily life" changed dramatically in the last quarter of the nineteenth century, and were increasingly developed—although steadily debated—by the early years of the twentieth. In this period, educators serving constituencies ranging from professionalizing engineers to Americanizing reformers established a role in schools for programs they variously called manual training, hand training, trades training, and industrial education. Not simply technological, these educational programs entwined ideologies about the social categories into which students might be sorted with ideologies about the technological categories into which different curricula might fall. But they were not simply ideological, either: institutionalized training provided access—or limited it—to particular forms of technological knowledge, knowledge that could shape opportunities in the student's future.

Examining both technological content and ideological justification of educational programs helps us map the "color line" in technological knowledge from around 1890 through the early twentieth century. But even preliminary examination of industrial education in this light reveals a twisted boundary. Whereas Coppin's Industrial Department had sought to provide access to something more than the fringes of the urban industrial economy, a decade later the ICY's "industrial school" in the country did nothing of the sort. The new curriculum stressed woodwork and basic metalwork, and the students were employed first in making the new quarters function as a working farm. Tuskegee, in contrast, despite its reputation, taught much more than simple agricultural science and basic rural trades. The Institute's African-American faculty offered instruction in mechanical and architectural drawing, and the workings of steam engines and electric plants, along with the more widely publicized training in brickmaking, horseshoeing, and hoeing. Booker T. Washington managed to keep these more obviously industry-related and "high tech" offerings—which were fully in keeping with his emphasis on economic self-improvement—at least tolerated in Alabama, and heavily supported and substantially financed by Northern industrial philanthropists. By the early twentieth century simple woodworking appeared regularly throughout the United States in curricula aimed at the lower echelons of American youth; running electrical plants, however, did not.

Thus the color line in technological knowledge must be traced through the national and regional tangle of attitudes toward race and toward science, technology, and industry. This work involves understanding ideologies of technology as well as of race and other social categories, looking closely at the meanings of words like "industrial" in different contexts, and interrogating our assumptions about American regionalism. Just as the color line in other realms was drawn differently in the Jim Crow South and the industrial North, so in the case of technological knowledge must we be aware of regional patterns.

The work presented here suggests that the paradoxes of industrial education spring from—and must be explained in terms of—contradictions between the various perceived potentials of the African-American labor force in New South industrial development; the tradition of large-scale production through hand cultivation in Southern agriculture; the steady technological marginalization of the African-American community in the urban North; and Northern philanthropy's funding of educational programs in ongoing conversation with Southern leaders. Understandings both of race and of technology varied regionally and in contested ways in the reconstructive decades following the end of "Reconstruction." Studying education carefully offers a window onto the connections between social categories and technological ones. This essay provides some tools for exploring the content and meanings of technical and vocational education, and then looks more closely at the details of Industrial Education at the Institute for Colored Youth, Hampton, and Tuskegee. A preliminary investigation, it invites a more complex portrayal not only of Industrial Education, but of the entwined categories described by "race" and "technology" in this period, in both South and North.[7]

Ideologies of Technology

Differences in ideologies of technology and technological knowledge—the meanings of teaching particular activities even under a generic rubric like industrial education—can be readily seen by comparing programs under the related umbrella "manual training" offered by the public schools of a single Northern city in this period. Philadelphia, home of the ICY, was a bustling manufacturing city boasting a broad range of industry and an ongoing interplay of progressive reform, machine politics, U.S.-born and immigrant residents, and an old and well-established African-American community. The public schools by no means provided the whole of the technical education or even the manual training offerings in the city, as the case of the privately funded ICY suggests, but a sketch of various public school curricula illustrates connections between social and technological categories. Such connections and shades of meaning are crucial to understanding the workings of industrial education, and to the study of race and technology in non-educational contexts.

The national Manual Training "movement" was initially a high school based initiative, intended to provide future engineers and industrial managers with the technological knowledge and the tools of what Edward Stevens has called "technical literacy"—mathematics, mechanical drawing, basic science—background they might need to master the most sophisticated production technologies of their era.[8] Philadelphia's Central Manual Training School (CMTS), a three-year high school program, was founded in 1885. But "manual training" was soon adapted to the perceived needs of other groups of students, and in 1890 the Board of Public Education opened an experimental Elementary Manual Training School in a neighborhood of recent immigrants, because "a school in which elementary manual training should be combined with primary school work, in the English language, would benefit the community by helping these children to a better life."[9]

To attend CMTS, the manual training high school, a boy needed to complete grammar school, score well on an entrance exam, and of course come from a family who could spare his wages for three years. By the senior year, his class project involved construction of "a steam engine, a dynamo, or some other machine," beginning with design and continuing with the making of working drawings, and progressing through the forging and machining of parts.[10] The purpose of such a project, educators noted, was to show the boys that they were prepared to do the work, and to demonstrate to them that even when prepared, new problems requiring thought could always arise.[11] Indeed proponents of this kind of manual training had often to argue that it was not trades training, but a scientific and problem-solving education for men who were "thinkers and doers." As they put it, "the shop instruction is simply part of the laboratory methods of education"; the school intended "not to make mechanics, but to train boys for manhood."[12] The curriculum at CMTS outside of shopwork and drawing included standard high school coursework in English, history, various sciences, and modern languages.

In Philadelphia's Elementary Manual Training School (the James Forten School), conversely, the children of a dwindling black community and a growing immigrant neighborhood were taught to sew pictures on punched cards, to cut and fold paper, and eventually to make simple wooden objects with hand tools or (for girls) to prepare food "plainly and economically."[13] In these cases the students were taught to copy the model provided by the teacher, to follow directions, rather than designing projects themselves. Educators boasting about the methods of the Forten School claimed this curriculum taught "habits of accuracy, neatness, dispatch, and obedience."[14] Not expected to stay in school past the grammar grades, the children at Forten needed, as city elders saw it, to learn American ways and factory obedience. Their curriculum, like that of the aspiring engineers and managers at CMTS, was labeled "manual training." But Forten students learned neither general mastery of science and technology, nor specific marketable trades.

Other programs in the Philadelphia schools further illustrate the connections between social ideologies and ideologies of technology, demanding analysis of gendered meanings and practices in addition to race and class distinctions. In the 1880s the Girls' Normal School was the only available public education for girls beyond the grammar schools. In 1881, at the joint direction of the Board's committees on the Normal and on Industrial Arts, teachers at the Normal added a course in sewing. Sewing was not yet part of the public school curriculum in any other school, so the point was not to prepare teachers for classroom work; the manual training programs discussed above had not yet been introduced, and no other public school taught any kind of hand work other than drawing. But as the Committees reporting on the idea asserted, "we believe that all women should have a knowledge of the art of sewing, whether they do or do not turn that knowledge to practical uses."[15] Sewing —an "industrial art" by Board definition—was appropriate to high school girls just as following directions was appropriate to impoverished immigrant children, and mastering design to young men preparing for life through mechanical drawing and modern languages.[16]

Thus the Institute for Colored Youth was developing its Industrial Department (opened 1889) and Booker T. Washington was establishing Tuskegee (founded 1881) at the same time other educators throughout the United States were developing patterns for school-based technical education. Hampton's programs in this period relied increasingly on trades work, as two-thirds of its students attended the Night School (working by day) by 1890.[17] Fanny J. Coppin's purpose of direct economic benefit by learning specific trade skills would not be echoed in the public school system for some years, but connections like those at Forten between respectable morality and some form of technical education were recognized by ICY managers and added to ICY's academic curriculum—and had been well established at Hampton from its founding in 1868. Thus while Coppin worked for economic opportunity, the encroachment of various forms of morally valuable but far less lucrative technical training on the academic program at ICY was steady in the 1890s, until the managers reinvented the school entirely at the turn of the century.

The Institute for Colored Youth

By 1880, Fanny Jackson (later Fanny Jackson Coppin), energetic Principal of the ICY, knew that African-Americans were under-represented in Philadelphia industry and craft occupations, and believed that access to formal industrial education would improve access to better jobs. An inveterate and indefatigable community activist, she began a campaign to add an industrial component at the ICY, hoping to expand the purposes of the institution.[18]

In this quest she was initially at odds with the (white) managers of the Institute. In 1882, when the income from a fund left expressly for the purpose of scientific education became available, she hoped the time for an industrial department had come.[19] The managers felt sufficient pressure, apparently, to address the issue in their annual report, but they were not enthusiastic. "The subject," they explained, "of opening an Industrial School in connection with the Institute was brought under consideration, but the Managers did not feel prepared to authorize such a step at this time. The income from the fund, indeed, would scarcely be adequate for the purpose."[20] There was no mention of the project in the annual report of the following year.

By 1884, however, the managers had begun to shift positions, and both rhetorical and curricular change was underway at ICY. The meanings of "industrial" work shifted too. The tasks of housework, already the province of African-American females, were the first to be taught in the school: classes in cooking and sewing had been made available to "some of the pupils," outside of the coursework, that year. In practice, this addition made no difference in the population the school served, nor did it transgress the economic barriers of craft exclusion. Justification for the inclusion of industrial arts in the ICY curriculum remained grounded in teacher training rather than manufactures, and relied on maintenance of social roles rather than their redefinition.

The new cooking and sewing classes, in the managers' view, would serve the purposes of home and classroom:

> Both classes are now being carried on with encouraging results, and it is hoped that the teaching will be a permanent benefit to [the students], in enabling them better to perform their household duties in after life.
>
> It will also enable our Normal classes to give some instruction in their schools upon the important arts of the household.[21]

The managers could make such a statement boldly despite contemporary debates about curricula, because the contents of the subjects African-American teachers should teach to African-American children, especially with the opening of more and more "Negro schools" in the South, had undergone dramatic revision. With the growing influence of what has been characterized as the Hampton/Tuskegee model of industrial education, tool use and housework were appropriate subjects for colored elementary school children.[22] The place of industry in the classroom also informed the managers' explanation of why they had "decided to open a branch for the education of boys in the Industrial Arts": "Again we believe that the young men who are graduated as teachers from this school are not as fully prepared for their work as they would be if they had been instructed in some useful trade, or at least knew the use of tools."[23]

These changes could be labeled plainly. Both because such institutions were private rather than public and because they provided for African-American education, the elaborate rhetoric justifying programs at the James Forten Elementary Manual Training School was apparently unnecessary. Hampton's program, now well established, provided the backdrop for such discussions. Even so, the plan to open an Industrial Department as a separate "branch" meant the expansion of the Institute's purpose, and the possibility of addressing Fanny Jackson Coppin's concerns about trade training for the larger community.

Five years later, in 1889, the managers' increasing commitment to the industrial training project had finally resulted in the more substantial financial undertaking of a new Industrial School House on a newly purchased lot adjacent to the Institute. In preparation for their new Industrial Department, the managers had visited other schools, including Hampton, and had hired teachers—the first white teachers in ICY's existence as a literary school.[24] Although in 1888 they had discussed the project in frank economic terms, including which trades could be learned "quickly and economically" and preferring "such branches as are most in demand and can be carried on with the least capital and risk,"[25] by the following year—and after their visit to Hampton—the managers' prose had acquired a loftier moral tone, very much in keeping with what they would have heard from Hampton's aging but active founder and principal, Samuel Armstrong.[26]

Whereas the previous year the managers had written of "preparing many of the colored people to become good mechanics," the 1889 report placed economic considerations second: the managers "look[ed] forward with a good deal of hope to the future success of the school in enabling the colored people to raise themselves in the scale of dignity and self-respect which accompany the acquisition of the power of supporting themselves."[27] They had "started classes in carpentry, bricklaying and shoemaking for the boys, and in dressmaking, millinery, and cooking for the girls."[28] This report also mentioned student interest in other trades, such as printing, plumbing, and blacksmithing. The Board explained that while they wished to include any workable options, they "at the same time wish[ed] to avoid entering inconsiderately upon any experiments which might be unsuccessful."[29] While potential students wanted access to a range of technological knowledge, the managers' caution suggests their different understanding of what kind of education colored people might need.

Despite the managers' reservations, printing and plastering were added the following year, and the annual report was printed by the printing class. Even so, demand far outstripped available offerings. In the first year, according to the annual report for 1889, 120 students, male and female, were enrolled in the classes of the industrial department, and 325 more waited for admission. Applicants expressed interest in more than 15 different trades, only five of which were offered.[30] A year later, the department was teaching 203 students, and the Superintendent could report that 522 applications for classes had been received from males alone since the department had opened.[31]

The successes recounted in Coppin's Superintendent's Report in 1891 demonstrate clearly why so many students were interested in the new offerings. The oyster shucker's story, introduced above, was among the most successful; he had even hired other students from the class to work with him. Another student, formerly a laborer, reported making $3.50 in a day's work.[32] The managers, however, chose their words differently. They did not discuss the popularity of the school in terms of the difference such wages made in paying the bills; rather, they told their readers, "it should cause satisfaction and gratitude that at the present time about 500 colored children are receiving literary and industrial education free of charge, getting their minds enlightened and hands trained to earn an honest livelihood and thus become useful and respected citizens."[33] Whether such phrasing reflects the views of managers or of donors, it implies that the issues at hand were honesty and respectability rather than economic well-being. The students sought access to better pay by means of vocational training; from the perspective of local white philanthropy, however, industrial education for colored students was increasingly associated with morality.

Meanwhile, significant changes had been made in ICY's literary curriculum, as well. Each year the annual report listed the curriculum of the five years leading to graduation; in 1880, students learned geography, arithmetic, grammar, spelling, history, penmanship, and drawing in the first year, and progressed to the U.S. constitution, vocabulary, mental and moral philosophy, theory of teaching, "Evidences of Christianity," astronomy, zoology, and Cicero (read in Latin). In 1890, the standard printed course of study was similar except that it included topics under the heading "sewing." In the first year, students learned sewing "from hemming to all the stitches used on plain garments"; in the second through fourth years they moved on to "plain sewing and cutting out garments"; in the senior year they were learning dressmaking, and cutting and fitting by measurement. The senior year also included cooking. By 1891, the early sewing curriculum had been further differentiated. In the first year students learned "hemming, felling, reversible seams, patching, darning, button-hole making, and marking"; in the second, tucking, gathering, herring bone, and feather stitch; in the next two cutting out and making garments; and in the senior year dressmaking.[34]

We can assume that dressmaking was taught only to ICY's female students, but sewing carried different implications for colored students training to be teachers than for public school girls. As of 1890 at ICY, the report explained, "sewing, darning, and patching are taught to all the pupils, both boys and girls." It further noted, "Boys' Specialty—Sewing on Buttons, Patching, Darning, and Button-hole making." No detailed explanation is provided, but the managers pointed out that graduates could give instruction in these subjects.[35] Whether they were male or female, sewing was an asset in teachers of colored children. Thus, while in general the gender rules of manual training and industrial education were conservative in their assumption of female domesticity, with universal sewing the rules bent away from white norms. Had the boys learned tailoring, they might have retained a badge of manliness; basic mending, in the ideology of late nineteenth century technology, would have been unthinkably emasculating for white youth.[36]

Additional elements of industrial education were incorporated into the Normal School program—alongside studies of Latin and physics—with the introduction of a "kitchen garden" in 1889. The kitchen garden, a rough parallel of "kindergarten" rather than a vegetable-growing enterprise, was a means of teaching housework, a formalized educational playhouse within the school setting. At ICY, Normal students mastered its routines so that they could teach younger children. The origins of this method bear further investigation, but its purposes were not simply those of "domestic science." At the annual public exhibition held at the school in 1891, the managers reported, "all were much delighted with the precision of the systematic drill in the various branches of household duty."[37]

The conformity connoted by "systematic drill" resembles the conformity of manual training models at Forten far more than it does the autonomy of Latin composition, but Latin prose composition had been dropped from the ICY curriculum in 1890. "Systematic drill" also connotes the obedience and control of military discipline, which was employed explicitly at Hampton and Tuskegee, where male students marched in uniform from dorm to dining hall.[38]

Unlike Philadelphia's CMTS, the addition of technical education for ICY's most advanced students carried conflicting messages. Increasingly in this period, the educational meanings of "high school" and "colored education" were inherently incongruous. In the Industrial Department, on the other hand, curricular content provided an avenue of access to technological knowledge, and thus to work the students would otherwise have been denied. White carpenters, bricklayers, or shoemakers did not generally attend school to learn their trade; other tradesmen shared techniques on the job. Potential African-American artisans had to learn their trades by different means. For

Coppin and the community she sought to serve, locating technical education in a *school* was an innovative, and for a time highly effective, urban economic strategy.

Had the managers been seeking simply to shift from academic to economically pragmatic training when they reassessed the program on Coppin's retirement, they could have expanded the model already in place in the ICY Industrial Department. By 1901, the Industrial Department employed nine teachers, one each in carpentry, bricklaying, shoemaking, printing, tailoring, shorthand and typewriting, dressmaking, cooking, and millinery. Enrollment had increased to 276 students, and the report from the Industrial Department included its share of success stories: several printers worked at their trade, including two in the "colored printing office, who do all the printing of the Colored Odd Fellows in the State of Pennsylvania," several who started businesses for themselves, such as the furniture repairman who had "sufficient work to keep him constantly employed," and one teaching shoemaking at a colored school.[39]

But in the managers' assessment, "the development of new ideals in education demanded a radical departure from the course of more recent years," and Coppin's vision was not one they shared.[40] Coppin had not been one to synchronize her efforts with the workings of Booker T. Washington's Tuskegee "machine," maintaining an emphasis on literary and scientific education for the high school in keeping with her own training at Oberlin, along with the more "self-help" style emphasis outlined above on economic opportunity in the Industrial Department. Booker T. Washington, by the turn of the century, had developed a range of strategies to promote his version of race education to Northern philanthropists, not infrequently including a critique of other people's work. After Coppin helped reorganize a Friends' (Quakers) colored school in Virginia, Washington was critical: "There were almost no signs of thorough sweeping, dusting, scrubbing, keeping up repairs, etc. These are the lesson[s] that our people need."[41] Such writings approach self-parody, but the ubiquitous and influential Washington won over many Pennsylvania Friends interested in education, ICY managers included. Thus the managers now argued that training good colored teachers (whether they came from urban Philadelphia or rural Virginia, presumably) demanded a boarding school, "for giving practical daily lessons in household economy and management and in proper home living." The managers closed the urban high school and the Industrial Department, purchasing a farm of 116 acres at Cheney, about 20 miles from the city. In the new school, with an entirely new staff, they would be able to "coordinate the normal, industrial, and agricultural features with home life," thereby contributing to the "primary object of training teachers who can give industrial education to all the pupils who come under their care."[42] These teachers would work in the rural South, not the public schools of cities like Philadelphia.

The Hampton-Tuskegee Model at Cheney

The ICY managers had visited Hampton in the late 1880s at the beginning of their foray into technical education. Samuel C Armstrong, white founder of Hampton, was the obvious person to consult, and Hampton was the model of success to observe. At Hampton, religious training and the importance of hard work and rigid discipline were combined with basic literacy in an educational program designed to "improve the race." As Armstrong put it in 1884, "not mere ignorance but deficiency of character is the chief difficulty." He believed that "a routine of industrious habit . . . is to character what the foundation is to the pyramid."[43] Hampton's program emphasized agriculture and Christian morals more than trades under Armstrong's leadership, although increasing numbers of students worked during the day to pay their way and attended the Night School rather than the Normal School within Hampton. Hollis B. Frissell, former chaplain of Hampton, served as President from 1893 to 1917, a period in which "the idea of trades training for economic purposes grew more influential nationally, and was at Hampton blended with the older model of character training."[44]

By the end of the 1890s, however, the better-known leader and spokesman for industrial education was not Frissell, but Armstrong's protégé Booker T. Washington. Thus the ICY managers paid a visit to the Tuskegee Institute as they contemplated the direction of the school in the new century. By 1901, the Annual Report explained the importance of donor support by noting the role of ICY graduates in the South: "Practical education that makes the race more self-respecting and more capable of maintaining this self-respect, by becoming useful members of society, seems to promise not only success for the present but security for the future. Money invested in this way yields a very tangible, present return and gives promise of a larger future harvest, as such centres of education multiply through our graduates in the South."[45] When Coppin retired, they followed Washington's advice and hired Hugh Browne, Principal of the Colored High and Training School, Baltimore, and former teacher at Hampton.[46] Also at Washington's request, they offered a summer school for Southern teachers, and they made regular use of his name and Margaret Murray Washington's in Annual Reports in the decade after the move to Cheney. Apparently, the managers not only looked to Tuskegee for ideas, but assumed that Washington's name would inspire donors in the mid-Atlantic region. Despite the ICY's half-century at the heart of Philadelphia's own African-American community, the managers focused their attention increasingly on the southern United States.[47]

Nonetheless, the ICY at Cheney—later renamed the Cheney Training School for Teachers—did not become a carbon copy either of Tuskegee or of Hampton. As discussed above, the retrenchment at ICY did mean a steady erosion of the academic curriculum until the managers essentially dismantled it. But simultaneously, the move meant a reduction of the trades-oriented training of the Industrial Department, because the new curriculum aimed at teacher preparation. As we shall see, in this same period both Hampton and Tuskegee developed more extensive trades school programs, but these facets of the Hampton-Tuskegee model were not generally exported northward. By 1911, after some years of experimentation, the ICY managers had found a niche for their institution, which they believed to be "one of the most intelligently planned and hopeful institutions to-day for the solution of the Negro problem and for the benefit of the race." As they explained, their "special ambition at Cheney is to develop a typical Normal School of such merit that it will be a pattern for state Normal Schools for Colored Youth throughout the South."[48]

Thus prospective teachers attending Cheney learned to teach the kinds of manual training and domestic science becoming standard in public schools nationwide, but as Browne and the managers developed the curriculum, Cheney ceased teaching trades. The severely curtailed academic program reflected assumptions about the levels of literary education colored students—and thus their teachers, and the teachers of those teachers—would need. By 1911, the faculty included three academic teachers with college degrees, and seven others, one educated at the college level. (See table 1.)

By 1911 the three-year course for teaching domestic science included 307.5 hours of science, 115 hours of education, 105 hours of art, and 534 hours of "Science Applied."

Table 1
ICY faculty, 1911. Source: ICY AR11, pp. 4–5.

Teacher	Education	Subjects taught
Hugh Browne	Howard	Applied physics and general methods
Naomi Spencer	Atlanta	History and allied subjects
Evangeline Hall	Radcliffe	English and mathematics
Gladys Caution	State Normal, Hyannis, Mass.	Physiology, hygiene, gymnastics
Laura Wheeler	High school + Pennsylvania Academy of Fine Arts	Drawing
Lewis Comegys	High school + ICY	Wood working
George Conway	Hampton	Iron working
Mabel Moorman	ICY	Domestic art
Julia Phillips	Columbia Teachers College	Domestic science
Ada Saulsbury	ICY	Applied domestic science

Science Applied included "all the forms of practical housework, namely: the need, selection, purchase, preservation, preparation, and serving of food; disposal of waste, utility of leftovers, and care of house." Listed as "the subjects of definite instruction" were "cookery, dietetics, marketing, serving, household economics, food production and manufacture, laundry work and dairying."

The course for teaching manual training included 802 hours of instruction in a range of techniques: woodworking, paper folding, sloyd,[49] venetian iron work, mechanical drawing, cardboard construction, and basketry. Mechanical drawing included the making of working drawings, most likely in preparation for other classroom activities. Thus graduates of these programs were in general prepared to earn their livelihood by teaching but not by trade work; the jobs of graduates listed in the annual reports by this time were almost all in education.[50] By contrast, the same decade had been one of further development of trades training at Hampton and Tuskegee. Although the ICY managers had, at the outset, sought to remake the school in the industrial education mold, as they proceeded they moved away from training workers and opted instead to train teachers, carefully prepared "for the benefit of the race." The degree to which this shift was discussed with or encouraged by Booker T. Washington is worthy of further investigation. Tuskegee, in this same period, shifted emphasis away from teacher training.

The Hampton-Tuskegee Model at Hampton and Tuskegee

In focusing on teacher training, the ICY managers may have been modeling their institution directly on the public presentations of Hampton and Tuskegee, but they were not following the practice of either school. Both Hampton and Tuskegee emphasized agriculture and, increasingly, trades training in this period—industrial education for livelihoods other than teaching. A comparison of programs and attitudes among the schools raises larger questions about who imagined which roles for African-American workers in the industrializing South, and about the how various regional constituencies viewed the racial character of modernity and civilization in this period.

At Hampton, these decades saw a general shift from Armstrong's original intentions. Thus Armstrong reported to his fellow alumni at Williams College in 1874 that the product of a Hampton education was a gentleman: "I have a remarkable machine for the elevation of our colored brethren, on which I mean to take out a patent. Put in a raw plantation darkey and he comes out a gentleman of the nineteenth century. Our problem is how to skip three centuries in the line of development and to atone for the injustice of the ages."[51]

A Hampton publication in 1912, aimed at Southerners rather than New Englanders, updated both the technological metaphor and the production goals of the institution:

The institution is a crucible—a veritable crucible—in which for nearly half a century a modern miracle has taken place. From out the South, where Negro life has been an existence of mental and moral darkness, steeped in ignorance, idleness, and superstition, a steady ceaseless current has flowed into this crucible, there to be transformed by the wonder-touch, and in turn poured forth, a living stream of moral integrity, mental strength, and industrial ability.[52]

In this vision electricity and steel, with an appropriately amenable work force, would provide the infrastructure of the New South.

Creating viable trades programs and selling African-American modernity, even mixed with moral development, was no simple matter in either Virginia or Alabama in this period, but increasing emphasis on trades training is nonetheless evident in both programs. Agricultural education also remained central, but at Tuskegee the normal program received less attention, except in the continuing summer schools for rural teachers. Further, beginning in the 1890s after Armstrong's death and gathering steam after the turn of the century, Tuskegee was no longer following the lead of Hampton.[53] Instead, Washington operated here as in other spheres: publicly he appeased and accommodated, while privately he both consolidated his own power and worked to expand opportunity for African-Americans.[54]

Washington and his supporters readily engaged in a convenient blurring of Hampton and Tuskegee, but hyphenating "Hampton-Tuskegee" or referring to the "Hampton idea" masked several important differences between the two. Most notable (and well known to supporters) was the all-Negro faculty, staff, and administration at Tuskegee, which was not a feature at Hampton. This circumstance was referred to condescendingly by visitors at times: "It is wonderful how much Dr. Washington has been able to accomplish with the assistance of those of his own race."[55] But there must have been a clear difference in the message each institution sent to students, many of whom came from rural schools scraped together by impoverished communities. The Washingtons, a black family, lived in a large Victorian home boasting modern plumbing, electric lighting, and a regular rotation of student servers at table; several of the faculty were college graduates, and they dressed with the care befitting salaried professionals, employing the services of the dressmaking, millinery, and tailoring departments.[56] For white visitors, too, a community in which all ranks of the hierarchy were filled by people of color was a new experience, striking for some, perhaps unsettling for others. At Hampton, for example, a suggestion in 1889 by African-American members of the staff that dining arrangements might be integrated was met with resistance and refusal by white teachers.[57] Thus no matter how anyone—incoming student or curious visitor—might understand Washington's words when he was traveling, a visit to Tuskegee offered material evidence that respectable middle-class morals and lifestyle might be achieved by a man working "Up from Slavery." Hampton preached this message; the Washingtons and the Tuskegee faculty lived it.

Differences between the two institutions, moreover, went beyond the racial composition of the faculty. Visitors also noted—in both positive and negative terms—a basic difference in spirit between Hampton and Tuskegee. Under Washington, evidently, Tuskegee was shifting from Armstrong's moral and religious teachings to an industrial emphasis arguably well in line with the more economic "self-help" focus Washington regularly set out as his prescription for race improvement.[58] As one observer noted, the word "institute" suited Tuskegee better than "school"—it was more flexible. "Here the industries dominate and color everything else, and the institution reflects both the excellence and the defects of an industrial community."[59]

Such an emphasis did not, however, align well with the ideologies linking simple techniques with happy plantation singing, immaculate schoolrooms, and other antidotes to immorality and idleness.[60] By 1906, several observers reporting to the Rockefeller Foundation's General Education Board (GEB) compared Hampton to Tuskegee and found Tuskegee wanting in spiritual emphasis. Washington had "become a victim of the American ideal of bigness," reported one visitor, suggesting the trustees should enforce limits on the growth of the institution. This policy would "have as one of its direct results an increased attention to the moral and spiritual conceptions that should inspire the life of the school." He elaborated the point, with reference to Hampton's Hollis Frissell:

When I think of the religious life of Hampton under the inspiration of a quiet, simple, but powerful man, of whom I cannot think without a thrill of admiration, and contrast that life with the dominant life of Tuskegee, I realize how almost infinitely greater and more promising for the future is the spirit which grows and deepens at Hampton, partly as a result of the great wisdom involved in its self-limitation.[61]

Another observer echoed the concern, weaving religion, discipline, and intelligence deftly together:

In respect of thorough correlation o[f] theory and practice and strong ethical and religious tone, Tuskegee is at a disadvantage compared with Hampton. That perfect discipline, alertness of application and intelligent grasp of subject matter as seen in all phases of the work at Hampton are not evident at Tuskegee.

These condescending critics expected Tuskegee to reach "higher results . . . in due time."[62]

But evidence suggests that Booker T. Washington and the staff he had assembled by the turn of the century did not aspire to quite those same "higher results," despite their willingness to justify their work in whatever terms seemed most likely to sell. Washington regularly explained in the catalog that under slavery, every plantation was a kind of trades school; without slavery a trades school was needed so his people could continue to fill the constructive role in Southern society that had always been theirs. This argument was cleverly designed to appeal to the nostalgia and regionalism of white Southerners while offering Northerners a vision of race and industrialization alternative to the exclusions of the North. Washington invoked the plantation past but hired R. R. Taylor, an MIT trained architect, to teach mechanical drawing, run campus industries, and design new buildings as needed. Taylor's notes in the Washington Papers document the struggles of running a rapidly growing "industrial community." In 1903, for example, he needed more "regular work students" in the machine shop, noting that the problem would get worse when the new boilers were set. There was often no engineer in the carpenter shop, he worried, from 10:30 to 12 daily, and there was no "walking engineer" assigned that year to report on various stations. He felt a walking engineer should also be assigned to the plumbing department, especially in case of a freeze, but there were not enough men assigned to that division either.[63] Further, an impending shortage of shoes suggested mechanizing shoemaking, or assigning more students to the shop.[64] In 1906 he reported that the industrial departments were now requiring overalls, and many students were obtaining them, but these students were having trouble getting them washed. The solution, of course, would be an extension of the laundry department and the use of more machinery there.[65]

Reading such arguments with attention to technological activity does indeed suggest that Washington had caught the American spirit of bigness, or at least of material and technological enthusiasm. As W. T. B. Williams told the GEB in 1906, presumably with reference to the lack of religious spirit others had lamented:

There is a spirit at Tuskegee as deep and dominating as one will find anywhere. But it is a spirit of getting on in the world, of material well being and thrift. This is taught six days in the week and often preached on Sunday. It naturally colors everything. Tuskegee is primarily concerned with laying that material foundation upon which alone a people may successfully rear their future civilization. An industrial atmosphere pervades the place and its air is more that of an industrial community than that of a conventional school. There is something of the hustle and bustle of the business world.[66]

That Tuskegee provided a painfully limited literary education is well documented. But as a working demonstration of the potential Negro role in the New South, it may have been much more effective, despite the chronic difficulties such a project entailed. The "Wizard of Tuskegee" was fully capable of echoing Hampton humbleness, but his creation in Alabama drew on industrial rather than Christian values.

The extent to which the Northern philanthropists funding the project were aware of and supported the differences between Hampton and Tuskegee deserves further attention. Northern views of race and civilization tended to exclude racialized peoples from full access to the progress of the age. David Nye has argued that electricity, as the "high tech" of this period, was culturally linked with both civilization and whiteness. Carolyn Marvin found that race was one of the codes for electrical outsider, along with women and rural rubes who failed to understand lighting or telephones, trying, for example, to insert written messages into the phone or to blow out a light bulb.[67] And this racial exclusion from artisanal and engineering practice was material as well as symbolic; Reverend Matthew Anderson, principal of the Berean Trades School in Philadelphia, emphasized that even with trades training, "only the more humble positions, such as those in ordinary day labor and waiting have been open to the colored man."[68] Washington's Tuskegee thus transgressed the racialized rules of progress and the primitive, whether or not the philanthropists intended that result. And Washington could make use of these rules, as well: the inherent racism of many visitors and auditors seem to have allowed him to engage in another of his "trickster" manipulations, using the excuse of inexperience to mask expenditures he thought might not align with his donors' ideas about what he should be doing. The records of the GEB, a substantial donor well-represented on Tuskegee's Board of Trustees, include repeated discussions of Tuskegee finances and bookkeeping techniques (as we shall see below).

Nonetheless, Tuskegee was no secret, and some basic material features of Tuskegee were well known. For example, the magazine *Technical World* (later continued as *Popular Mechanics*) ran an article in 1904 by a Tuskegee teacher, Charles Pierce, on "How Electricity Is Taught at Tuskegee." This article explains that the course was begun in 1898.[69] By way of comparison, the ICY at Cheney, near cosmopolitan Philadelphia rather than in rural Alabama, briefly offered "applied electricity" a decade later, and in the Annual Reports of 1910 and 1911 was printing appeals justifying the advantages of "an electric lighting plant to replace the unsatisfactory and dangerous oil lamps."[70] Electricity had not been part of the physical plant of a Pennsylvania farm school in 1902, but Tuskegee was already alight. And several Tuskegee spin-off schools scattered across the rural South had electrical plants before World War I, at a time when fewer than 5% of households in states like Alabama or South Carolina had electricity.[71]

Washington and his faculty borrowed arguments for Tuskegee's programs from his audience, of course, and deployed them cleverly. Pierce's *Technical World* article begins by explaining the great danger of fire when oil lamps were used, on account of the carelessness of students in handling the lamps, before going on to detail the equipment, curriculum, and skills of Tuskegee electricians, and to mention that Tuskegee's telephone exchange meant the school employed "the first colored 'hello girl' (to the writer's knowledge) in the world." Pierce does mention that the Institute supplied electricity to several places "off the school grounds, including the residence of the late . . . Congressman from this district," which may explain the apparent local tolerance of cutting-edge technology. In his final paragraphs Pierce again mitigates any admiration the reader might have acquired, perhaps increasing his readers' tolerance for the technological advancement of a colored school, by invoking the tropes of electrical outsiders. He recounts stories of the visiting farmer who wanted nothing to do with the "dynamite machine" in the Dynamo Room, the man who wanted to know how oil could pass through a solid wire, and the student who thought he had finally succeeded in blowing out the lamp in his room only because the dynamo was shut down each night. And Pierce concluded with a veiled threat of idleness as the alternative to technological knowledge: carefully trained Tuskegee students had no trouble finding work, demonstrating that Tuskegee was "teaching the dignity of labor and of learning to do things well."

Perhaps such disarming strategies misled the gentlemen of the GEB somewhat. By 1905 they do seem to have been trying to explain to themselves the differences in expenditures between Tuskegee and Hampton. (Table 2 gives the totals from a GEB tally of 1904–05.) Hampton's overall budget was larger, but Tuskegee's total for "Industrial Departments and Trade School" was nearly three times Hampton's, while the Academic Group spent little more than half Hampton's total. Washington argued, meanwhile,

Table 2
Hampton and Tuskegee expenses, 1904–05. Source: Comparative Statement, Hampton Institute vs. Tuskegee Institute, Year 1904–1905, folder 24, box 3, series 1.1, General Education Board Archives, RAC.

	Hampton	Tuskegee
Administration and General Expenses	$69,541.47	$39,215.38
Academic Group	64,746.58	33,790.58
Raising Funds Group	8567.13	12,520.27
Maintenance Group	29,007.95	33,596.43
Extension, Outside Expense Exhibits, etc.	6837.19	9697.96
Indust. Departments and Trade School	18,066.60	52,632.61
Sundry Trading Dept [incl. boarding dept]	5353.33	2388.19
Total of above	$202,120.25	$183,841.42
[with further unexplained adjustments]	$211,078.80	$190,978.72

that hiring former students for work on campus out of the educational budget made sense because it was really "post-graduate" training.[72]

Nonetheless, there was a general acceptance of the program. Seth Low, considering the chairmanship of the board of trustees in 1907, sought to make sure his ideas and Washington's were in accord. In his view, Tuskegee needed fewer students, and higher standards. "I am fully in sympathy with your idea of placing the emphasis at Tuskegee's work on the industrial side; though, naturally, I should like to see the academic side made at the same time as strong as possible. But it is evidently quite as important that the students of Tuskegee should have a good reputation industrially and academically; and it is impossible for that result to be obtained unless the students stay at Tuskegee long enough to be moulded by Tuskegee." He understood, he told Washington, the desire to give a little to many, but believed that showing the effects of a thorough Negro education by Negroes for Negroes was also important.[73] His comments suggest an alternative to Washington's strategy for justifying his programs.

Unquestionably there were problems preventing the "thorough education" of students, and not only those occasioned by Washington's ambitions. The economic climate of the South in these years offered choices to students, as Williams's report pointed out: "To my mind the most serious hindrance to the teaching of trades thoroughly to the masses of the students lies in the difficulty in keeping the students long enough. The South is greatly in need of skilled mechanics. As soon as these boys get a fair knowledge of their trades they can find profitable employment. They take this in vacation time and fail to return to finish their courses." Thus, Williams argued, the incomplete preparation of some students—obviously a concern of the GEB—was not the school's fault. On the whole, he reported, Tuskegee's program was successful: "The number of well-trained men turned out is gratifyingly large. The school has no difficulty in placing them and the men of this class are giving satisfaction."[74] This report stands in noticeable contrast to the comments Berean's Reverend Anderson offered about Philadelphia's labor market segmentation.

Finally, in exploring the meanings of industrial education programs, we must recognize that educators' ideas about technical education in this period were dynamic, and as complex as the racial, class, and gender ideologies of their day. While some situated the importance of Tuskegee's program firmly in the economic development of the New South, and others in the improvement of the race, still others saw a developing facet of pedagogical theory. Paul Monroe of Columbia Teachers' College explained his views to the GEB and is worth quoting at some length:

My interest in Tuskegee and a few other similar institutions is founded on the fact that here I find illustrated the two most marked tendencies in present educational endeavor; tendencies which are being formulated in the most advanced educational thought, but are being worked out slowly and with great difficulty. These tendencies are first, *the endeavor to draw the subject matter of education,* or the "stuff" of school room work *directly from the life of the pupils*; and second, *to relate the outcome of education to the life's activities, occupations, and duties of the pupil* in such a way that the connection is made directly and immediately between school room work and the other activities of the person being educated. This is the ideal at Tuskegee, and, to a much greater extent than in any other institution I know, the practice; so that the institution is working along not only the lines of practical endeavor, but of the most advanced educational thought.[75]

As noted in the discussion of Philadelphia's public schools above, such "advanced educational thought" assumed that educators could predict the "life's occupations" of students, whether children of immigrant families in New York or rural families in Alabama. That the plumbing, electricity, and mechanized shops of Tuskegee bore no resemblance whatsoever to the prior "life of the pupils" in the rural South does not seem to have concerned Professor Monroe. It did, however, concern the denizens of Tuskegee, for whom such difference in lifestyle was very much part of the point.

Somewhat ironically, despite their ostensible interest in occupations, white educational theorists were in general not interested in—and were at times either oblivious or perhaps antagonistic to—the economic advantages of practicing trades as an "outcome of education." Samuel Dutton, also of Columbia's Teachers College, demonstrated his vision of student futures in his comment on visiting the dressmaking and millinery departments: "I found that these industries are not taught except to those who choose them as special trades. . . . The head teacher [in millinery] defends the practice on the ground that she is a member of the union and that it would not be right to give away the secrets of the institution." Dutton disagreed, contending that such trades should not be secret but should be regarded as "universal things": ". . . every girl should have a few practical lessons with such reference to those economies which such people ought to practice, so that when they go out they can carry into their homes and schools the simple ideas." Finally, Dutton cited Hampton's principal to underscore his point: "While all these universal activities may properly be taught as trades, they are so essential in the homes and schools of the South, that I think every student should have them. I found that Dr. Frissell agreed with me in this opinion."[76] Of course, "simple ideas" were not likely to draw customers to a millinery business, but Dutton does not seem to have espoused the goal of economic opportunity and mobility. He disapproved also the idea of teachers wearing hats purchased from the millinery department.

In fact the economic and educational goals at Tuskegee, notwithstanding the praise of some educators, *were* sometimes at odds. As in any industrial production in the United States at this time, sound business principles demanded division of labor, not conceptual understanding. Thus Margaret Murray Washington reported to her husband in June 1903 that the millinery department had been focusing on the "business side," making "salable articles" rather than doing graded work. The emphasis was not ideal for beginners, but the seniors were all making hats for commencement now. The girls assigned to plain sewing had been making caps and aprons, and bedding, and cash orders were steadily filled. Work on a large order of shirts, night shirts, and underwear was underway. She saw a difficulty despite the general success in the dressmaking department: "Although there have never been as many orders filled for expensive dresses of teachers and white patrons, nor a better designer in charge of the Dressmaking Division, the lack of business tact and teaching ability on the part of the instructor is a heavy drag. I recommend an assistant who will have charge of the making of uniform dresses and other simple ones, and teach the class in drafting."[77]

On the boys' side as well, reports suggest that the balance between teaching, completing work, and making money was hard to achieve, both within the industrial department and between the academic and industrial departments. W. T. B. Williams, generally supportive of the industrial programs, reported: "There is general complaint among the industrial teachers that their pupils are not well enough prepared to take their trades to the best advantage. It not infrequently happens that the "rule of thumb" methods must be taught the pupils because they are not advanced enough in academic work to comprehend other methods." Williams suggested that certain classes could be required as a prerequisite to training in some trades, rather than assigning all students to begin trades as they entered school. "This too would be in keeping with Tuskegee's avowed ideal—industrial efficiency rather than general intelligence dissociated from the utilities."[78]

While such problems were apparently chronic, and they have been highlighted by other historians, it is worth noting also the apparently accepted goal of "industrial efficiency." On the surface it seems no surprise that a collection of Northern philanthropists including the likes of Rockefeller and Carnegie might support such training, but Northern industry did not typically hire black industrial labor.[79] The knowledge and especially the wages of industrial work were opportunities to which the African-American community of Northern cities had trouble getting access.

Region and Race: Ideologies of Technology

The question raised by such comparisons, then, is about the regional shape of ideologies of technology. The standard Northern story about the exclusion of African-Americans from industrial work contrasted with the simultaneous story of Northern philanthropy encouraging the possibility of black industrial labor in the South. Certainly the ideal of industrial efficiency was not uncontested—reports from visitors contain repeated pleas for religion and morals. Indeed this public version of the Hampton-Tuskegee program was ubiquitous and was often the message carried northward, but clearly some Northern industrialists, possibly less publicly, were persuaded that black workers could fuel Southern industry. Possibly they assumed that new Southern industrial workers would stay Southern, and perhaps remain more tractable than their troublesome white Northern counterparts. Northern philanthropic attitudes, and the attitudes of white Southern educators and progressives, bear further investigation.

What seems clear from this exploration, however, is that both racial and technological ideologies were entwined in the industrial creation of the New South, that these ideologies cannot be treated as nationally uniform, and that regional influence was fully reciprocal. The contours of contested ideologies—the possibility of training cheap labor for growing American industry or as docile agricultural workers, the provision of industrial or manual training for mastery of a trade or for obedience in a factory—have yet to be studied carefully. The details of Southern industrial whiteness, and white technical education, must be examined as carefully. These issues need explaining because electrical plants and steam engines in the rural South cannot be dismissed as automatic corollaries to limited literary preparation. The paradoxes of industrial education spring from—and must be explained in terms of—contradictions between the logical role and perceived potential of the African-American labor force in New South industrial development, on the one hand, and on the other, the steady technological marginalization of the African-American community in the urban North.

Of course, as with other accommodationist strategies, Washington's emphasis on industrial efficiency came with the cost of enhancing someone else's power: Tuskegee students learned to follow directions and obey rules, to focus on pay rather than worker control. Where Coppin's institutional route to technological knowledge was clearly intended to subvert the social and technological order—the exclusion of her community from trades work—Washington's played to the industrialists. Still, in an era when "civilization" was both racialized and explicitly technological, the very fact of an industrial community, no matter how problematic, challenged not only the oppressive economics of the South, but also, implicitly, the racial and technological order of the North. Ironically, of course, Washington was only too successful in mitigating the effects of such challenges: every time he argued that "his people" needed to learn to scrub floors while remaining silent about his dynamos, he disseminated northward a technological as well as a moral caricature of the rural South.

We have tended too easily to view Industrial Education as a monolithic and standardized educational alternative to what might be called the Black College model, treating it as an unexamined "other," a foil rather than a phenomenon in need of explanation. Examining its various meanings more closely raises questions about the racial and technological dimensions of nation-building in the decades after Reconstruction. Neither the educational discussions nor the social and economic implications of these curricula belong to just one region: as the preceding discussion suggests, one extended network of educators and philanthropists could support a Southern school with a distinctly Northern industrial flavor and a Northern school preparing teachers for the rural South. These contradictions were characteristic not only of the Tuskegee Machine or of the New South, but also of a New North even before the larger migrations of the coming decades.

Acknowledgments

This work was supported in part by the National Science Foundation and is indebted to Bruce Sinclair, Lynette Cook-Francis, Carol Harrison, Julie Johnson-McGrath, Christian Gelzer, Paul Lerman, Annelise Heinz, Deborah Douglas, Amy Slaton, and Andrea Gass, to Carroll Pursell and the History Department at Case Western Reserve University, to the Whitman College History Department, and to Cynthia Wilson at the Tuskegee University archives.

Notes

1. Institute for Colored Youth, *Annual Report for 1891* (hereafter ICY-AR91), p. 18. For limits on economic opportunity for African-Americans in this period see W. E. B. Du Bois, *The Philadelphia Negro: A Social Study* (published for the University of Pennsylvania, 1899); Roger Lane, *Roots of Violence in Black Philadelphia, 1860–1900* (Harvard University Press, 1986); Theodore Hershberg, ed., *Philadelphia: Work, Space, Family, and Group Experience in the Nineteenth Century: Essays toward an Interdisciplinary History of the City* (Oxford University Press, 1981).

2. The curriculum included Latin, English, history, moral philosophy, physics, chemistry, astronomy, and zoology. The science department was headed by Edward Bouchet, Ph.D. Yale 1876. ICY ARs 1880–1902.

3. On Coppin, see her *Reminiscences of School Life and Hints on Teaching* (Phildelphia: African Methodist Episcopal Book Concern, 1913; reprint Hall, 1995) and Linda Marie Perkins, "Quaker Beneficence and Black Control: the Institute for Colored Youth, 1852–1903," in *New Perspectives on Black Educational History*, ed. V. Franklin and J. Anderson (Hall, 1978). See also Coppin's Principal's Reports in the ICY Annual Reports.

4. See James Anderson, *The Education of Blacks in the South, 1860–1935* (University of North Carolina Press, 1988); Donald Spivey, *Schooling for the New Slavery: Black Industrial Education, 1868–1915* (Greenwood, 1978). On Booker T. Washington see Louis R. Harlan's two-volume biography, *Booker T. Washington: The Making of a Black Leader, 1856–1901* (Oxford University Press, 1972) and *Booker T. Washington: The Wizard of Tuskegee, 1901–1915* (Oxford University Press, 1983). Anderson's otherwise excellent book is weak on the industrial side of the curriculum (for example, on p. 76 a photo of a technical drawing class is mislabeled as an English class). Spivey did not have the benefit of Harlan's second volume.

5. Robert Engs's excellent biography of Samuel Chapman Armstrong highlights the mid-century meaning of "industry" in Armstrong's design of the Hampton curriculum. Robert Francis Engs, *Educating the Disfranchised and Disinherited: Samuel Chapman Armstrong and Hampton Institute, 1839–1893* (University of Tennessee Press, 1999). By the 1880s, terms like "industrial arts" designated topics like mechanical drawing; the Philadelphia Board of Public Education had a Committee on Industrial Arts by 1881. See *Journal of the Board of Public Education* (1881).

6. Anderson, *The Education of Blacks in the South*.

7. The importance of the national framework of regional ideas about race, and in particular the neglect of Northern racism, is an issue for all U.S. historians. For the revolutionary and early national periods see Gary B. Nash, *Race and Revolution* (Madison House, 1990). I have also been influenced by the following works: David Roediger, *Wages of Whiteness: Race and the Making of the American Working Class* (Verso, 1991); Glenda Gilmore, *Gender and Jim Crow: Women and the Politics of White Supremacy in North Carolina, 1896–1920* (University of North Carolina Press, 1996); Gail Bederman, *Manliness and Civilization: a Cultural History of Gender and Race in the United States, 1880–1917* (University of Chicago Press, 1995).

8. For an explanation of the phrase "technical literacy," see Edward W. Stevens, *The Grammar of the Machine: Technical Literacy and Early Industrial Expansion in the United States* (Yale University Press, 1995). For a more extended discussion of manual training, see Nina Lerman, "The Uses of Useful Knowledge: Science, Technology, and Social Boundaries in an Industrializing City," *Osiris* 12 (1997): 39–59.

9. Philadelphia Board of Public Education, *Annual Report for 1891* (hereafter BPE-AR 1891), p. 17.

10. Catalogue of the Manual Training Schools 1890–91, p. 23.

11. Calvin Woodward's address is quoted in Charles A. Bennett, *History of Manual and Industrial Education, 1870–1917* (Charles. A. Bennett, 1937), p. 358.

12. *Catalogue of the Central Manual Training School, 1894–95* (Philadelphia: Board of Public Education, 1894), p. 29; *Catalogue of the Manual Training Schools 1890–91*, n.p.

13. BPE-AR 1892, p. 137. The Forten School was originally a public "colored school" within the city school system, but by 1890 changing demographics led the BPE to reinvent the school. Presumably without irony they kept the name of the successful African-American sailmaker whose early-nineteenth-century artisanal wealth earned him a place on the roster of local emulation.

14. BPE-AR 1891, p. 126.

15. *Journal of the Board of Public Education* (1881), appendix 48.

16. It is important to note both the gender *inclusiveness* of terms like "manual training" and "industrial arts," and the gender *differentiation* (usually) of content under those universal rubrics.

17. Engs, *Educating the Disfranchised and Disinherited*, p. 157. Engs finds Samuel Chapman Armstrong caught between donor expectations, pragmatic economic solutions, and the ideals he developed in founding the institution—which were thus less clearly implemented in Armstrong's final years.

18. Coppin, *Reminiscences*; Perkins, "Quaker Beneficence and Black Control."

19. For Fanny Jackson Coppin's views see Perkins, "Quaker Beneficence and Black Control"; Coppin, *Reminiscences*; Harry Silcox, "The Search by Blacks for Employment and Opportunity: Industrial Education in Philadelphia," *Pennsylvania Heritage* 4 (1977), December, pp. 41–43. Silcox's forceful portrayal of the interests of the African-American community unfortunately suffers from history of education's habitual blindness to the many meanings of terms like "manual training" and "industrial education."

20. ICY-AR82, p.14. Note that "science" versus "industrial" education was not the issue.

21. ICY-AR84, p. 12.

22. For background on Hampton and Tuskegee, see Anderson, *The Education of Blacks in the South*; Engs, *Educating the Disfranchised and Disinherited*; Vincent P. Franklin and James D. Anderson, eds., *New Perspectives on Black Educational History* (Hall, 1978); Donald Spivey, *Schooling for the New Slavery: Black Industrial Education, 1868–1915* (Greenwood, 1978); Louis Harlan's biography of Booker T. Washington. Tools and housework had long since been considered appropriate training for some children, as at Houses of Refuge (reform schools) or schools for the deaf. See Nina E. Lerman, "Uses of Useful Knowledge" and "'Preparing for the Duties and Practical Business of Life': Technological Knowledge and Social Structure in Mid-Nineteenth-Century Philadelphia," *Technology and Culture* 38 (1997): 31–59.

23. ICY-AR84, p. 13.

24. ICY-AR89, pp. 4, 12–14.

25. ICY-AR88, pp. 11–12.

26. On Armstrong's views of morality and uplift, and on his ability to shape his argument to his audience, see Engs, *Educating the Disfranchised and Disinherited*.

27. ICY-AR88, p. 14; ICY-AR89, pp. 13–14.

28. ICY-AR89, p. 13.

29. ICY-AR89, p. 14.

30. ICY-AR89, pp. 12, 17–20. The most frequently requested trades other than the ones offered at ICY were shorthand, typewriting, plumbing, printing, and upholstering.

31. ICY-AR90, pp. 15–16.

32. ICY-AR91, p. 18. The oyster shucker's $7/week for a 6-day week would likely have meant $1.17/day.

33. ICY-AR90, p. 17.

34. ICY-AR80, pp. 6–8; ICY-AR90, pp. 8–10; ICY-AR91, pp. 10–12. ICY discussions of sewing mention paper patterns, which were more prevalent by late century, but not sewing machines. Cooking was taught only to girls.

35. ICY-AR90, pp. 10, 13.

36. At the House of Refuge, delinquent boys did learn tailoring. But it was always described as such and was performed sitting cross-legged on top of the tables in the time-honored position of tailors. Lerman, "Uses of Useful Knowledge."

37. ICY AR91 p. 16

38. Armstrong was a General in the Civil War and used the title throughout his life. See Engs, *Educating the Disfranchised and Disinherited*, pp. 34–35. But a similar discipline was instituted at the Philadelphia House of Refuge, a reform school. See Philadelphia House of Refuge, Annual Report for 1892, p. 32. Photographs from the 1890s show boys in uniforms marching in orderly formation into or out of the schoolhouse and other buildings.

39. ICY AR01, p. 17.

40. ICY AR03, p. 10.

41. Booker T. Washington (hereafter BTW) to Ellis Perot Morris, March 30, 1896, in *The Papers of Booker T. Washington*, ed. L. Harlan (University of Illinois Press), vol. 4, pp. 150–151.

42. ICY AR03, pp. 10–14.

43. Samuel Chapman Armstrong, *Lessons from the Hawaiian Islands* (1884) quoted in Engs, *Educating the Disfranchised and Disinherited*, p. 74

44. Donal F. Lindsey, *Indians at Hampton Institute, 1877–1923* (University of Illinois Press, 1995), pp. xiv, 247; Harlan, ed., *Booker T. Washington*, vol. 1.

45. ICY AR01, pp. 18–19.

46. ICY AR02, p. 16. Browne was a graduate of Howard University and Princeton Theological Seminary. He was head of physics at the M Street High School and then at Hampton. *The Papers of Booker T. Washington*, vol. 5, p. 50, n. 5. His ideas about the ICY at Cheney bear further investigation.

47. For example, James Dillard of the Jeanes Fund, a philanthropy dedicated to Southern education, visited Cheney and is quoted in ICY AR10.

48. ICY AR11, p. 35.

49. Sloyd (elementary woodworking based on a Swedish system) was also taught at the James Forten Elementary Manual Training School in Philadelphia.

50. ICY AR11. Women's technological knowledge was generally more marketable; one woman was employed in dressmaking.

51. Quoted in Engs, *Educating the Disfranchised and Disinherited*, p. 152.

52. J. W. Church, *The Crucible: A Southerner's Impression of Hampton* (Press of Hampton Normal and Agricultural Institute, 1912). The role of electricity in the civilizing mission here was not idiosyncratic; see David Nye, *Electrifying America: The Social Meanings of a New Technology* (MIT Press, 1990).

53. Several reasons contributed to the shift: Armstrong died in 1893; Washington was better known after his Atlanta speech in 1895 and gathered power; Tuskegee's legal status changed so private donations became possible, rapidly eclipsing state funding.

54. See Harlan's portrayal of Washington, especially in volume 2. Harlan did not look at industrial and technological issues. See also Barbara Bair, "Though Justice Sleeps, 1880–1900," in *To Make Our World Anew*, ed. R. Kelley and E. Lewis (Oxford University Press, 2000).

55. Samuel Dutton, Teachers College Columbia, to Robert Ogden, April 16, 1906. folder 25, box 3, series 1.1, General Education Board Archives, RAC.

56. This pattern is most evident in complaints by visitors but can also be found in sales records. Further exploration of the cash economy centered at Tuskegee would be worthwhile.

57. See Engs, *Educating the Disfranchised and Disinherited*, pp. 161–162.

58. For further discussion of Tuskegee and both Booker T and Margaret Murray Washington, see Bair, "Though Justice Sleeps," and Gilmore, *Gender and Jim Crow*.

59. W. T. B. Williams, "Confidential to members of the General Education Board. Tuskegee Institute. WTB Williams, May 1906," folder 49, box 6, series 1.1, General Education Board Archives, RAC.

60. The role of music also warrants further investigation. On Armstrong's uses of music at Hampton, see Engs, *Educating the Disfranchised and Disinherited*.

61. Samuel Bishop to Ogden, March 23, 1906. folder 25, box 3, series 1.1, General Education Board Archives, RAC.

62. Samuel Dutton (Teachers College, Columbia) to Ogden, April 16, 1906, folder 25, box 3, series 1.1, General Education Board Archives, RAC.

63. R. R. Taylor (RRT) to BTW, December 29, 1903, Cont 591 reel 441 Industries 1903, Booker T. Washington Papers, Library of Congress.

64. RRT to BTW, December 9, 1903, Cont 591 reel 441 Industries 1903, Booker T. Washington Papers, Library of Congress.

65. RRT to BTW Mar 14 1906, Cont 615 reel 452, Booker T. Washington Papers, Library of Congress.

66. W. T. B. Williams, "Confidential to members of the General Education Board. Tuskegee Institute. WTB Williams, May 1906," folder 49, box 6, series 1.1, General Education Board Archives, RAC, p. 36.

67. David Nye, *Electrifying America*. On civilization and race, see especially Bederman, *Manliness and Civilization*; see also Michael Adas, *Machines as the Measure of Men: Science, Technology, and Ideologies of Western Dominance* (Cornell University Press, 1989); Carolyn Marvin, *When Old Technologies Were New: Thinking about Electric Communication in the Late Nineteenth Century* (Oxford University Press, 1988), chapter 1.

68. Anderson was pastor of Berean Presbyterian as well. Rev. Matthew Anderson, DD, "The Berean School of Philadelphia and the Industrial Efficiency of the Negro," *Annals of the American Academy of Political and Social Science* 33 (1909), January, p. 113. He goes on (p. 114): "Would that this country and the leaders of our great commercial interests could be made to see the golden opportunity that presents itself to them. Twelve millions of people are a mighty host!"

69. Charles W. Pierce, "How Electricity is Taught at Tuskegee," *Technical World* 2 (1904): 425–431.

70. ICY AR1910, pp. 17–18; ICY AR1911 p. 34.

71. See William H. Holtzclaw, *The Black Man's Burden* (Neale, 1915), p. 231, on the Utica [Mississippi] Normal and Industrial Institute; J. F. B. Coleman, *Tuskegee to Voorhees: The Booker T. Washington Idea Projected by Elizabeth Evelyn Wright* (Bryan, 1922), p. 126, on the

Voorhees [SC] Normal and Industrial School. On rural electrification, see Nye, *Electrifying America*, chapter 7. For further discussion of the early days of electricity, see Marvin, *When Old Technologies Were New*.

72. See BTW reply on accounting, folders 25 and 26, box 3, series 1.1, General Education Board Archives, RAC.

73. Seth Low to BTW, May 29, 1907, cont 934 reel 691, Booker T. Washington Papers, Library of Congress.

74. W. T. B. Williams, "Confidential to members of the General Education Board. Tuskegee Institute. WTB Williams, May 1906," folder 49, box 6, series 1.1, General Education Board Archives, RAC.

75. Paul Monroe, Teachers College, Columbia University, to Wallace Buttrick, May 6, 1904,, folder 24, box 3, series 1.1, General Education Board Archives, RAC (emphasis added).

76. Samuel Dutton, Teachers College Columbia, to R. Ogden, April 16, 1906, folder 25, box 3, series 1.1, General Education Board Archives, RAC.

77. Mrs. Washington to Mr. Washington ["June 1903" written in pencil] Cont 591 reel 441 Industries 1903, Booker T. Washington Papers, Library of Congress.

78. W. T. B. Williams, "Confidential to members of the General Education Board. Tuskegee Institute. WTB Williams, May 1906." folder 49, box 6, series 1.1, General Education Board Archives, RAC.

79. For general discussions and further references, see Mary Frances Berry and John W. Blassingame, *Long Memory: The Black Experience in America* (Oxford University Press, 1982); Kelley and Lewis, eds., *To Make Our World Anew*.

Raising Fish with a Song: Technology, Chanteys, and African-Americans in the Atlantic Menhaden Fishery

Barbara Garrity-Blake

Work is at one time an economic, political, and religious act and is experienced as such.
—Maurice Godelier[1]

In considering the role of technology for post-bellum Southerners who left field for factory, we usually think of capitalist-controlled machines that brought about increased profits for industrialists but declines in skill, control, and autonomy for black and white laborers.[2] The Greek root *techne* means art, or craft, but most often we attribute the genius and creativity in technology to the inventors of machines, not to what artisans themselves brought to the process of getting things done. Workers, particularly African-Americans in the historical period of wage labor and segregation, tend to be viewed as cogs in the wheel of industrialism, having little input beyond the muscle power used to keep the machines fed and humming.

The Western bias toward defining technology in terms of machines and tools limits our understanding of alternative means to industrial ends; it masks the importance of less tangible forms of technics, such as laborers' styles of working, the accumulation of knowledge in the workplace as situations demand, and the will to solve problems and complete tasks.[3] A narrow "machines and their inventors" understanding of technology furthermore minimizes the creative and intellectual contributions of non-capital controlling laborers; the myriad of techniques working people have devised to do work, sometimes *in spite* of the machines or ideas of superiors, are too often overlooked.

This essay examines the role of expressive culture among African-American crewmen of the Atlantic fish meal and oil industry who sang work songs during harvesting operations for much of the twentieth century. The work song or chantey was more than an accompaniment to the work process: it was used by laborers as a necessary means of raising the heaviest sets of fish.[4] The collective act of singing, in other words, allowed Southern black fishermen to perform feats they would not otherwise have been able to do, and it demonstrates that critical technologies do not necessarily come out of formalized learning or from the use of mechanical devices.

If we take cultural understandings of the work process seriously, the significance of laborers' contributions in American history becomes more complex and central.[5] Our assumptions about the nature of labor, class, technics, and progress are challenged. In this case, where crewmen brought to the work setting a method fashioned by the unique historical experiences of African-Americans in the South, the U.S. menhaden industry was ultimately dependent on the power of their crews' own intangible technology: the creativity, spirituality, and shared strength generated in singing.

The Menhaden Industry in the South

The Atlantic menhaden, commonly known as "mossbunker," "shad," "poggie," or "fatback," is one of the most prolific fish species in U.S. waters. Menhaden, however, has rarely been marketed as a food commodity because of its high oil and bone content.[6] By the turn of the century the menhaden industry, beginning in New England fifty years earlier, had replaced whaling as the largest commercial fishery and fish oil producer in the nation. In addition to oil, fish solids were dried and marketed as fertilizer.

After menhaden became scarce in northern waters toward the end of the nineteenth century, Yankee industrialists began moving their operations south to take advantage of fish-rich waters and, more importantly, the abundance of cheap labor in the post-war period of Reconstruction.[7] By 1912, the geographical center of the menhaden industry had shifted to North Carolina and Virginia, when thirty-one factories operated in these Southern states compared to less than half that number in the Northeast. By World War I a new menhaden product was marketed in addition to oil and guano: fish meal, a dried and powdery by-product sold as a high protein additive to livestock feed. Fish meal would come to be the "bread and butter" of the U.S. menhaden industry for the latter half of the twentieth century.[8]

Most menhaden companies were vertically integrated, and company owners controlled everything from the fishing operations, to the factory reduction of menhaden, and the final marketing of products. Factory equipment included raw boxes, cookers, screw presses, rotary driers, conveyers, and oil tanks. Nineteenth-century sailing vessels were replaced by coal-burning wooden steamers or converted Navy mine sweepers, each equipped with two cedar purse boats and a "striker" row boat. Purse nets were some 1,200 feet long, constructed of tarred and salted cotton twine.[9]

The labor-intensive menhaden fishery drew workers from small coastal villages to work the vessels, factories, net houses, mess halls, and company shipyards. Offering the first steady paycheck for many rural black and white Southerners in the early twenti-eth century, menhaden companies came to be known as the economic backbone in "fish factory" towns. And making a virtue of necessity, villagers in places such as Reedville, Virginia, and Beaufort, North Carolina, proclaimed the awful stench from processing plant smokestacks that blanketed their towns "the smell of money."[10]

Shipboard Working Conditions

Unlike most other U.S. fisheries, menhaden fishermen did not sell their catch to dis-tributors or processors, but were decidedly hired employees.[11] So these workers did not enjoy the quasi-independent status of fishermen who owned their own boats and oper-ated in a relationship of debt and patronage with a seafood dealer. Instead, the men-haden companies themselves generally owned fleets of the wooden steam-powered vessels, each steamer manned by crews of twenty-seven to thirty-four men. And since they all worked for the same firm, captain and crewmen were at least alike in their lack of control over the means of production.

 By 1920, however, the majority of labor for the U.S. menhaden industry fleet had changed from Yankee and Portuguese fishermen to Southern African-Americans, and that added yet another dimension to labor patterns. In Virginia and North Carolina, black laborers were primarily from land-owning farming and fishing families in out-lying rural communities.[12] They came to fill the bottom rungs of a rigid shipboard hier-archy, as the captain, mate, fireman, engineer, and pilot were typically white, while some two dozen deckhands and the cook were black. With the absence of kinship or ethnic ties commonly found between captains and crews of many other fisheries, men-haden crewmen and officers were decidedly unequal, an inequality understood both in terms of occupational role differences and Southern race relations.

    Like menhaden factory yards, the eating and sleeping facilities on most vessels were segregated according to race. Black fishermen slept below, in some cases directly under fish-covered decks, where the slime and brine dripped through the cracks and kept their bunks damp. Officers slept in more comfortable quarters, with the captain enjoying the most spacious cabin directly behind the wheel house of the vessel. It was not uncommon for ship's galleys to have had a "colored" section for taking meals. Captains carried firearms on board which, although for shooting sharks that occasionally thrashed in the net, further underscored the status differences between deckhands and higher officers.

    Ship captains recruited crew members for their strength and ability to endure the backbreaking task of menhaden fishing. Once the captain spotted fish from his perch in the crow's nest, the whole crew save the cook and pilot would board two small shallow-draft purse boats and, guided by a man in a smaller "striker boat" who signaled the movements of the fish, they rowed toward the school of menhaden. Upon reaching the fish, the boats separated with a tremendous purse seine net playing out between them. After encircling the fish, a heavy "tom" weight was dropped to draw the bottom of the net closed, purse-string fashion, and trap the menhaden inside. The floating, corked top of the net was clutched by crewmen and pulled back in the purseboats by the armful until the catch was "raised" and "hardened" for bailing into the mothership.

It is unclear when and where menhaden crewmen first began employing songs to raise fish, but witnesses reported singing in the waters off North Carolina by the 1920s.[13] Yet, except for those passing sailors and yachtsmen who had the good fortune of hearing chanteys roll out over Chesapeake Bay or Atlantic Ocean waves, few experienced the singing that occurred only in context of particularly heavy sets of fish.[14]

Like work songs in other contexts, the menhaden chantey was a means by which workers could coordinate their efforts. After the leader sang out a refrain and the crew responded in harmonious chorus (baritone, bass, alto, etc.), the men would "fall on the net" and haul in an armful or "fall back." Unlike tempo-setting chants of most black American worksongs, however, menhaden chanteys were characterized by a broken rhythm: between each verse the men encouraged, scolded, and cursed each other while pulling. The verses and chattering continued until the catch was concentrated near the water's surface.

Up to a half million fish would then be bailed into the hold of the mothership with a "dip net." The dip net was rigged with a counterweight and a block and tackle. One of the deckhands would climb the mast and secure himself in a boatswain's chair that was attached to the dip net ropes. Upon jumping, the force of the crewman's fall raised the fish-heavy dip net, allowing the crew to take in the slack. This was repeated dozens of times until the entire set was loaded. It was common for crewmen to make several sets a day. Typically, captains would not return to home port for days, until the vessel was sunk low with fish. If inclement weather did not overwhelm the dangerously heavy steamer, often arriving at the factory with its decks awash, the crew could look forward to good pay and a day or two of rest after unloading the fish, restocking the coal, and salting the nets.

Modeled after the whaling industry, the crew was paid according to the "share" or "lay" system: calculated on the basis of their shipboard status, the men received a percentage of the value of fish caught per trip. If fish were plentiful and the crew successfully loaded the steamer, paychecks would reflect the bounty. But if fish were scarce, weather was bad, or a heavy set of fish capsized the purseboats or split the net, it was conceivable that the crew would receive no pay at all for hours, days, or even weeks of work. Deckhands received the smallest share, ranging from one third to one fifth of what the captain received.[15]

Menhaden Chanteys

The origin of menhaden chanteys is best understood in the context of African-American history rather than the European maritime tradition of sea chanteys. Unlike the robust refrains heard for centuries aboard merchant ships, the menhaden chanteys are relatively slow and mournful, containing floating verses and themes noted in many non-maritime African-American contexts such as Antebellum plantations, post-war prison yards, tobacco factories, and railroad camps.[16]

African-American work songs are characterized by their improvisatory nature, as verses are mixed freely to the extent that the lyrics of discrete songs are difficult to pin down. A leader typically varied the song each time it was sung, borrowing from the "entire spectrum of traditional Afro-American songs, combining them with improvised lines."[17] Whether secular or spiritual, the work songs involved the West African-inspired style of "call and response" singing: the crew leader sang out a refrain, and the deckhands answered in "grand chorus" of several harmonies.

Like African-American laborers in other work contexts, menhaden fishermen could express discontent and resistance within earshot of the "boss man" through singing. The unifying theme of all the chanteys was a desire to be someplace else: on a new job, in a woman's arms, or in a state of grace with God. One of the few refrains that seem to be unique to menhaden fishing is found in "Sweet Rosie Anna": "steamboat coming 'round the bend, bye bye sweet Rosianna . . . I won't be home tomorrow . . . I won't be back till next payday." "Evalina" is about a disreputable woman with a "money accumulator" between her legs. Frustrations of working offshore with a crew of men rather than tending to the homefires are expressed literally in "the house is on fire, fire, fire, and it all go burning down." Some songs, such as "I Gotta Woman," clearly reflect desire:

I gotta woman
SHAKES LIKE JELLY ALL OVER
(repeat)
from the hips on down
LORD, LORD FROM THE HIPS ON DOWN
Last time I seen her
LEGS JUST AS BIG AS HER BODY
(repeat)
gonna rub 'em down, Lord
LORD, LORD, RUB 'EM DOWN[18]

Another prevalent theme reflected in menhaden chanteys is resistance to and rebellion against authority. Unlike their ancestors who may have worked as slaves or tenant farmers, twentieth-century menhaden fishermen were wage laborers, free to quit their employment at will. And unlike their captains, who so strongly identified with the occupation of fishing, deckhands made a sharp separation between who they were and what they did, having little qualms about changing boats, companies, or jobs. Several chanteys reminded the captain of this fact, such as "Dark Cloud," where the singers asked, "Cap'm have you heard all your men goin' to leave you? . . . On the next pay day." Just as the same images "wander through the various [African-American] worksong traditions, respecting neither sacred nor secular boundaries,", so did the laborers move "from occupation to occupation, ever adapting their songs to new environments."[19] Sentiments of resistance were closely related to ideas of solidarity between crewmen, who placed utmost value on "pulling together." "Cap'n if you fire me," one verse warned, "you got to fire my buddies too."

The African-American folk hero Lazarus, who stood up to the white power structure, is sung about in work contexts and prison yards throughout the American southeast. Menhaden fishermen sang that Lazarus "shit on the commissary counter" to protest low wages and company-owned stores, getting shot and killed for this act of defiance.

"Bitin' Spider," one of the most enigmatic chanteys, appears to describe a close encounter with an insect "going 'round bitin' everybody." While some crewmen interpreted bitin' spider as representing the captain, others said it was the fishing net that's "killed many a man." The net, with its "webbing" spread out over the water, has caused generations of torn muscles and ligaments, and inflamed arthritic fingers so badly that crewmen could not uncurl their hands at the day's end. The net was also known to entangle and drown fishermen after a capsize, or bring about heart attacks. It was thus a tool that could prove productive when catching fish but destructive in the hard work and risks of menhaden fishing. The refrain "but he didn't bite me, Lord" captures the ambivalence of working a job that could range from lucrative to deadly.[20]

Raising Fish with Song

Menhaden fishing represents one of the last contexts in which African-American worksongs were heard, continuing until the early 1960s when hydraulic net-lifting devices, or power blocks, became pervasive throughout the industry and rendered the singing, as well as half the labor force, unnecessary. But during the industry's most labor intensive period companies and captains were dependent on the sheer strength and endurance of the predominantly black crews. Although some captains explicitly recognized the central role of crewmen, and bragged on their singing and fishing abilities, others seemed to view crewmen as an extension of the equipment deployed in the fishing process. However, all captains claimed for themselves an intangible technology of their own, namely an almost mystical ability to "read signs in nature" and to spot schools of fish from cues as subtle as a certain hue of the water, or scent in the air, or even a dream from the night before. These talents defined successful captains, and they prided themselves on their ability to locate big catches of fish—believing this gift to be "in the blood" and therefore individualistic. But the ability to raise fish belonged to and was celebrated by African-American crewmen, for whom fishing was always a collective effort and the singing of chanteys a critical and necessary part of fishing operations.[21]

A key aspect of menhaden singing extends beyond providing a rhythm by which to work: the songs were reserved for the heaviest sets of fish. Crewmen placed enormous value on doing "everything by hand," and well-recognized the fact that they were hired primarily for their muscle power. For smaller sets, muscle power sufficed and fish were hardened unceremoniously, without song. But for sets containing two hundred thousand or more fish—more weight than muscle power alone could handle—crewmen turned to their own technique. One of the men would strike up a chantey—for example, "I've got a woman, LEGS JUST AS BIG AS HER BODY!"— and begin the ritualized process of raising the otherwise unmanageable load with a song[22]:

> With a heavy set it soon came to a point where muscle was not enough. . . . All knew there was work ahead to get [the fish] in the hold before they died to become dead weight, [so] inspiration came to the chanteyman. . . . Their voices, like their muscles, were attuned to everybody else's. The verses came out of the chanteyman's store of couplets built up over his years of fishing and hearing chanteymen before him.[23]

Some sets were so large that pulling the net risked capsizing the purse boats; captains would then signal a sister vessel, and enlist the aid of an additional crew who would lash their purse boats to the precariously tipping boats. Then some sixty men would "put a nickel in the jukebox" and commence to sing, bringing "heaven and earth" together and lifting an enormous quantity of menhaden.[24]

According to crewmen, heavy fish could be raised by song because singing in unison worked to generate a special "boost" or collective euphoria that masked the pain of the endeavor and served to conjure up superhuman strength: "You didn't know how much you was pulling. You'd be about happy there singing them songs, all them shads in that net . . . everyone feeling good and everything."[25]

So "blowing a chantey" enabled the men to raise what could not otherwise be raised by human strength alone. A certain focus and exhilaration was achieved whereby the men were "in tune," working as one with a new energy brought forth through singing.

Crewmen labored as if in a trance, all in one motion, eyes fixed on the "money" splashing around in the net. Singing, according to African-American laborers, is what put the "life" in menhaden fishing, lifting both their spirits and the fish that would enable them to go home and provide for their families.[26] And the power generated by singing reached beyond the vessel; in their recollections deckhands emphasized that wealthy yachtsmen and daysailors would anchor near the menhaden vessels to listen, held captive by the beautiful singing of laborers. "Nothing could make them move away from there," marveled a retired crewman, "till we were through singing that chantey."[27]

Technology and Southern Black Crewmen

In menhaden boats that seemed transformed into "a regular church on water," African-American laborers communicated a spiritual sense of rising above earthly hardships and transcending the work context of physical pain, danger, and low status.[28] The worksong as technology not only *functioned* as a tool to get the job done, it *signified*: the words expressed resistance to white authority, freedom to seek new wage-labor employment, and the desire to be home with loved ones.

The harmony of the songs reflected the importance of worker solidarity in enduring difficult work conditions in an uncertain and highly dangerous environment. Not unlike the state of "communitas" described by Victor Turner, intense "comradeship and egalitarianism" among crewmen was achieved by ritualistic means.[29] Success, for menhaden deckhands, depended not so much on the directives of company managers or captains, but on the mutual support of each other.

Although most of the profits from menhaden fishing went to company owners, and most of the credit for large catches of fish went to vessel captains, it was the deckhands who possessed the only means of raising impossibly heavy sets of fish, an awesome feat uniquely fashioned out of African-American expressive culture. The introduction of "power block" machinery, usually described as a pivotal point of progress for the fish-meal and oil industry, undercut the achievement of the crews. Before these hydraulic net-lifting devices silenced the song, the welfare of the highly industrialized and bureaucratic menhaden fishery rested not only on the strong backs of laborers but on the spiritual power of Southern black chanting.

Through the analytical category of invisible technology, the cultural significance of work becomes apparent, as does the unique and central role crafted by the "powerless," allowing us to reconsider the relationship between class and access to technology. Working for wages in early twentieth-century segregated America, menhaden crewmen lived with a system that was "not quite total," that left open "some if not many avenues of expression," and yet was "not only oppressive but ambiguous."[30] They enjoyed the choice and freedom of hiring on or quitting various jobs, but these were primarily low-status jobs in a white-dominated system that was not their own. What did belong to the African-American crewmen, however, was the technology of song that lay at the heart of the fishing process, with which they not only lifted the unliftable but expressed desires and discontents that were, if not unspeakable, risky to express for the times. Singing was thus a way black workers rendered their work culturally meaningful and effective in a context that otherwise devalued their contributions and racially divided them from their employers.

Appendix: "Lazarus"

Old man Lazarus
SHIT ON THE COMMISSARY COUNTER
Old man Lazarus
SHIT ON THE COMMISSARY COUNTER
Then he walked away, boys
LORD, LORD, HE WALKED AWAY

Captain told the high sheriff
GO AND BRING ME LAZARUS
Captain told the high sheriff
GO AND BRING ME LAZARUS
Bring him dead or alive, boys
LORD, LORD, DEAD OR ALIVE

Well he found poor Lazarus
WAY DOWN BETWEEN TWO MOUNTAINS
Well he found poor Lazarus
WAY DOWN BETWEEN TWO MOUNTAINS
He shot him down, boys
LORD, LORD, SHOT HIM DOWN

Oh Lazarus' mother
SHE CAME A SCREAMING AND A CRYING
Oh Lazarus' mother
SHE CAME SCREAMING AND A CRYING
Said, you shot my son, boys
LORD, LORD, YOU SHOT MY SON

Notes

1. Maurice Godelier, *Rationality and Irrationality in Economics* (NLB, 1972).

2. Harry Braverman, author of *Labor and Monopoly Capital: The Degradation of Work in the Twentieth Century* (Monthly Review Press, 1974), is a well-known proponent of the theory that

increased technology brought about declines in laborer skill, control, and autonomy in the twentieth century.

3. Judith A. McGaw, in her research on early American technology, calls for broadening the "tool- or machine-oriented" view of technology to include knowledge. *Early American Technology: Making and Doing Things from the Colonial Era to 1850* (University of North Carolina Press, 1994), p. 14.

4. Much of the research on the menhaden fishing process has either given scant attention to the chanteys or treated them as an interesting aside to the main discussion of companies, inventors, and mechanical innovations within the industry. Most attentive to the chanteys are folklorists. See, for example, Glenn Hinson, *Virginia Worksongs*, Blue Ridge Institute Record No. 007 (Ferrum College, 1993); and John Michael Luster, Help Me to Raise Them: The Menhaden Chanteymen of Beaufort, North Carolina, Ph.D. dissertation, University of Pennsylvania, 1994.

5. Herbert Gutman long ago called for studies examining the "beliefs and behavior of American working people" and how that "affected the development of the larger culture and society." *Work, Culture, and Society in Industrializing America* (Vintage, 1977), pp. xi–xii.

6. In the U.S. menhaden as a food fish has been associated with the underclass, and as a commodity salted menhaden was marketed in the nineteenth century to "serve as food for the negroes upon the plantations" of the West Indies and Guiana. George Brown Goode, *A History of the Menhaden* (Orange Judd, 1880), p. 135.

7. Goode, *A History of the Menhaden*, pp. 191–192; Robert N. McKenney, "The History of the Menhaden Industry in Virginia," *Northern Neck Historical Magazine* 10 (1960), p. 914.

8. Roger W. Harrison, "The Menhaden Industry," U.S. Department of Commerce, Investigative Report No. 1, 1931, p. 41.

9. Rob Leon Greer, The Menhaden Industry of the Atlantic Coast, U.S. Bureau of Fisheries document 811, 1915, pp. 6–10.

10. Barbara J. Garrity-Blake, *The Fish Factory: Work and Meaning for Black and White Fishermen of the American Menhaden Fishery* (University of Tennessee Press, 1994), pp. 15–27; John Frye, *The Men All Singing* (Downing, 1978), pp. 57–91.

11. Greer, *The Menhaden Industry of the United States*, p. 20; Ralph H. Gabriel, "Geographic Influences in the Development of the Menhaden Fishery on the Eastern Coast of the United States," *Geographic Review* 10 (1920): 95–96; Harrison, "The Menhaden Industry," p. 41.

12. Garrity-Blake, *The Fish Factory*, pp. 134–137.

13. Frye, *The Men All Singing*, p. 186.

14. In 1950 NBC sent a recording crew aboard the *Barnegat*, and for the first time a menhaden chantey, the spiritual "Drinking of the Wine," received national radio exposure. Hinson, *Virginia Worksongs*, p. 5.

15. Elmo Hohman, "Wages, Risks, and Profit in the Whaling Industry," *Quarterly Journal of Economics* 40 (1926): 644–671; Liguori, *Stability and Change in the Social Structure of Atlantic Coast Commercial Fisheries*, pp. 81–85.

16. Hinson, *Virginia Worksongs*, pp. 2–14.

17. Ibid., p. 7.

18. Library of Congress recording AFS 11,399. Words that are capitalized are those sung in chorus; normal typography indicates a solo voice.

19. Hinson, *Virginia Worksongs*, p. 14.

20. Garrity-Blake, *The Fish Factory*, pp. 106–107.

21. Ibid., p. 188.

22. Hinson, *Virginia Worksongs*, p. 13.

23. Frye, *The Men All Singing*, pp. 82–183

24. Garrity-Blake, *The Fish Factory*, p. 105.

25. Ibid.

26. Ibid., p. 103.

27. Ibid., p. 110.

28. Hinson, *Virginia Worksongs*, p. 23.

29. Victor Turner, *Forest of Symbols* (Cornell University Press, 1967), p. 95.

30. John W. Cell, *The Highest Stage of White Supremacy* (Cambridge University Press, 1982), p. 243.

Pictures from an Exposition

Bruce Sinclair

Among the U.S. exhibits at the Paris Exposition of 1900 was one called the "Negro Section," housed in the Palace of Social Economy. The person responsible for putting it together was Thomas J. Calloway, an employee of the War Department, who happened also to have been an undergraduate classmate of W. E. B. Du Bois at Fisk University. Indeed, Calloway asked Du Bois to help assemble materials for the exhibit, and Du Bois, who joined him in Paris, wrote about the experience afterwards in a piece called "The American Negro at Paris."[1]

The exhibit, a picture of which is shown here, included books, models, the patents of black inventors, information about instruction at Fisk, Howard, and Atlanta Universities, at Hampton and Tuskegee Institutes, and at a few other colleges devoted to the education of African-Americans. Besides these things, the display included hundreds of images that portrayed African-American men, women and children as respectable, middle class people: "photographs of typical Negro faces," Du Bois said, "which hardly square with conventional American ideas." That collection of photographs included, as Du Bois put it, "an especially excellent series," taken at Hampton Institute and "illustrating the Hampton idea of 'teaching by doing.'" Calloway had commissioned the noted woman photographer Frances Benjamin Johnson to make those pictures, an assignment that contrasts interestingly with her normal patronage by the rich and famous.[2]

In their self-conscious style the images may seem stilted to us, but Johnson (an iconoclast who clearly understood Calloway's intentions and the politics of her portrait-making) posed Hampton's students as attentive learners, eager for knowledge and disciplined in their pursuit of it. And those images stood in stark contrast to the depiction of African primitivism in the exhibits from France's colonial possessions. So besides reminding us of the value of photographs as historical source materials, the content of these samples from Paris reveal the crucial importance of self-representation for African-Americans, in family life, work, education, and—particularly to dispute the canard of disingenuity—in creative intellectual ability. Du Bois summed it up this way: "We have, thus, it may be seen, an honest, straightforward exhibit of a small nation of people, picturing their life and development without apology or gloss, and above all made by themselves."[3]

Calloway and Du Bois's exhibit, with Johnson's photographs, was recreated for the Pan American Exposition at Buffalo in 1901, mainly as a result of the efforts of the Phyllis Wheatley Club of Colored Women in Buffalo, and people thought it made an important statement there, too.[4] Many of the materials from that exhibition have long since been dispersed, but Eugene F. Provenzo, of the School of Education at the University of Miami, has recreated it on a web site.

Johnson continued to make pictures. In her later years she moved from Washington to New Orleans, where, as a free-spirited person, she enjoyed the city's easy conventions. Her photographs and papers are in the Library of Congress.

Notes

1. W. E. Burghardt Du Bois, "The American Negro at Paris," *American Monthly Review of Reviews* 22 (1900): 575–577.

2. Johnson's photographs at the Exposition and their importance are nicely analyzed by Jeannene M. Przyblyski in "American Visions at the Paris Exposition, 1900: Another Look at Frances Benjamin Johnson's Photographs," *Art Journal* 57 (1998), fall: 61–68. The most recent biography of Johnson is Bettina Berch, *The Woman Behind the Lens: The Life and Work of Frances Benjamin Johnson, 1864–1952* (University Press of Virginia, 2000). Also valuable is Pete Daniel and Raymond Smock, eds., *A Talent for Detail: The Photographs of Miss Frances Benjamin Johnson, 1889–1910* (Harmony, 1974).

3. Du Bois, "American Negro at Paris," p. 577

4. "Africans, Darkies and Negroes: Black Faces at the Pan American Exposition of 1901, Buffalo New York."

Figure 1
The "Negro Section" from the Paris Exposition of 1900.

Figure 2
W. E. B. Du Bois in Paris, for the Exposition of 1900. Courtesy of the W. E. B. Du Bois Library, University of Massachusetts and reproduced with permission.

Figure 3
"A class in capillary physics at Hampton Institute." Frances Benjamin Johnson Collection, Library of Congress, Prints and Photographs Division (LC-USZ62-108065). This caption and the ones that follow are the ones Johnson gave these pictures.

Figure 4
"Two young men training in mechanical (?) engineering at Hampton Institute." Frances Benjamin Johnson Collection, Library of Congress, Prints and Photographs Division (LC-USZ62-119865).

Figure 5
"Students at work on a house built largely by them." Frances Benjamin Johnson Collection, Library of Congress, Prints and Photographs Division (LC-USZ62-66946).

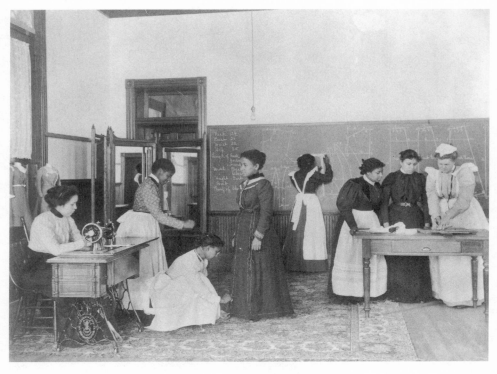

Figure 6
"A class in dressmaking, Hampton Institute." Frances Benjamin Johnson Collection, Library of Congress, Prints and Photographs Division (LC-USZ62-38151).

Figure 7
"Demonstration of milk testing in stable, at Hampton Institute." Frances Benjamin Johnson Collection, Library of Congress, Prints and Photographs Division (LC-USZ62-108067).

Figure 8
"Student shipbuilders at Newport News, Virginia." Frances Benjamin Johnson Collection, Library of Congress, Prints and Photographs Division (LC-USZ62-132387).

Figure 9
"Students constructing telephones at Hampton Institute." Frances Benjamin Johnson Collection, Library of Congress, Prints and Photographs Division (LC-USZ62-132386).

Figure 10
"Negro students working at shipbuilding at the Newport News Shipbuilding Co." Frances Benjamin Johnson Collection, Library of Congress, Prints and Photographs Division (LC-USZ62-66945).

"The Open Road": Automobility and Racial Uplift in the Interwar Years

Kathleen Franz

In July 1910 Jack Johnson, a black pugilist who had just defended his heavyweight boxing title against "White Hope" Jim Jeffries, posted a $5,000 challenge to the automobile racing world.[1] By his own account, Johnson was an "ardent motorist" who had owned cars since their first appearance in the United States.[2] Johnson offered his challenge in hopes of securing a contest with one or all of the top three drivers of the year: George Robertson, Ralph De Palma, and Barney Oldfield. Oldfield was the only one to accept, and he told the press that he was doing it not just for the money but to avenge Jim Jeffries. He had to accept Johnson's challenge "to stop all criticism and comment just as Jeffries had to fight him."[3]

Still smarting from Johnson's thorough defeat of Jeffries, civic officials, journalists, and racing authorities, such as the American Automobile Association Contest Board (which sanctioned and administered all official automobile contests in the United States), hesitated to frame the Johnson-Oldfield meet in racial terms. The boxing match had been promoted as a contest to prove the virility and authority of white manhood.[4] What if Johnson proved as superior a driver as he was a boxer? Would this mean that Johnson, and by extension African-American drivers generally, were capable of challenging white authority in terms of technological competency and spatial mobility?[5]

In the months preceding the contest, the white press focused on Johnson's traffic violations, portraying him as reckless and a threat to the general public.[6] At the same time, newspapers constructed Oldfield as a popular hero, dubbing him the "Speed King" because he had defied the rules of the AAA Contest Board. When Oldfield defeated Johnson on a rainy October 26, the white press observed that the superior skill of the driver, not the machine, had determined the contest. The *New York Herald* declared: "Oldfield Maintains Track Supremacy."

Jack Johnson was exceptional in his wealth and in his willingness to confront white authority, but his experiences as a motorist and the white press's representation of him as a poor driver typified the racialized nature of the automobile in the first third of twentieth century. Racial divisions were central to the construction of what historians of technology have termed automobility, or individual spatial mobility.[7] Racial discrimination on the open road and representations of African-Americans as technologically incompetent reinforced a belief in white superiority at a time when the white middle class was feeling the threat of cultural fragmentation and blacks had started gaining middle class status.[8] In the 1920s, white motorists relied on ideologies of racial difference established in the nineteenth century to reinforce the boundaries between the races on the open road.[9]

Johnson may have lost the contest, but he continued to take full advantage of both the social and the geographic mobility offered by the automobile.[10] As he and other members of the black middle class purchased the new machine and sought the freedom of movement it promised, they created a new site of racial contestation in the United States. For their part, black motorists developed a set of strategies that allowed them to access the new technology and its culture, and to challenge both segregation and white attempts to represent all blacks as unskilled drivers.

This essay investigates the automobile as an instrument of cultural power that offered black middle-class drivers access to both personal mobility and technological expertise. I argue that the emergent black middle class in the 1910s and the 1920s adopted the automobile not only as a means to circumvent segregation, but as a material expression of racial uplift ideology. This ideology held that African-Americans could assimilate into American culture through education, bourgeois morality, and a culture of self-improvement. The historian Kevin Gaines has written that "racial uplift ideals were offered as a form of cultural politics, in the hope that unsympathetic whites would relent and recognize the humanity of middle-class African-Americans, and their potential for . . . citizenship rights."[11] The concept of "race" progress in the early twentieth century focused on education and community action, but the ideology of uplift also had a material component, a cultural aesthetic that used material goods and, especially, the new technology of the automobile to demonstrate the achievement of a middle-class lifestyle.

Cultural histories of technology, with their focus on users, material culture, and consumption, suggest two ways in which we can expand the scope of African-American history.[12] First, a focus on material culture can help us broaden the study of the black middle class from the lives and ideas of intellectuals to the material practices of everyday life, such as consumption of new technologies and leisure travel. As historians of technology who study women have shown, consumption is a highly political form of technological use.[13] Second, studies of how the black community appropriated technological artifacts can highlight the importance of technological skill and knowledge within larger arguments about racial progress. As an example, I will show how black narratives of progress used the automobile to insert African-Americans into dominant discourses of economic prosperity, leisured mobility, and technological know-how.

On a Different Road: The African-American Middle Class

Between 1918 and 1939, automobility offered the small but growing black middle class an opportunity for greater spatial and social equality. Some scholars have asserted that blacks and whites used the automobile in similar ways, but most have emphasized the differences between the experiences of black and white motorists.[14] Sociologist Bart Landry, for example, noted that the composition of black and white middle classes differed. While the white middle class grew significantly between 1910 and 1920, the leap from manual labor to white-collar or professional work was more difficult for African-Americans, many of whom had rural backgrounds and fewer white-collar opportunities.[15] The black middle class of the 1920s differed from its white counterpart in education and in income; it comprised doctors, ministers, teachers, newspaper editors, and small businessmen, rather than clerical workers, salespeople, and managers.[16] Highly educated and increasingly self aware, this nascent black middle class sought full citizenship and social equality after World War I. Using service in the war as an example of national loyalty and capacity for citizenship, blacks sought all the material and social privileges of their class. These included, among other things, the freedom to purchase the new consumer goods that signified middle-class affluence and gave them greater access to public space.[17]

Car ownership among African-Americans grew more slowly during this period than it did among whites. Yet motor cars and the possibility of vacation travel became a material signifier of racial progress. Assessing the increase in automobile ownership among the black middle class in the 1920s, Robert Moton observed, "If motor cars are an index of prosperity there is significance in the fact that in every city of any considerable size Negroes are to be found in possession of some of the finest cars made in America."[18] More black Americans purchased automobiles after 1918 than they did before the war, partly as a result of the growth of the resale market.[19]

Black publications such as *The Crisis*, *Champion*, and *Opportunity* and newspapers such as the *California Eagle* promoted car ownership, provided vacation information, and defined the cultural meanings of automobility. Melnotte Wade, a contributor to *Champion*, encouraged her husband to buy a car in 1917, when "autos were the rarest of the rare in our community and no colored man had yet been the proud possessor of one." "I must confess," she wrote, "that it had long been the height of my ambition to travel some other way than by train or the monotonous jog-jog behind old Dobbins. . . . Shutting my eyes to all ideas of economy, I could see myself lolling back in a car of my own."[20] The remainder of the article was a warning to potential car buyers who gave in to such impulses. Mrs. Wade and her husband had been sold a lemon by an unscrupulous white car dealer. Nevertheless her dream of enjoying the freedom afforded by the automobile resonated with others in her community.

The call of the "free car and open road" resonated in middle-class black communities in the 1920s.[21] As Gaines has noted, the black elite shared the dominant culture's distaste for cities crowded with migrants from the rural South.[22] Over several weeks in June 1924 the *California Eagle* ran a series entitled "Where Will You Spend Your Vacation?" The *Eagle* suggested that its readers cure their vacation fever by taking a "vacation of fifteen days or more" consisting of a motor trip to the woods. The paper urged readers to cut costs by camping: ". . . we would advise that you rough it. . . . This affords you an advantage that is to be appreciated only by a visit to this wonderful place and too it reduces your expense to a minimum."[23]

In the same year, the pull of an outdoor motor excursion inspired five members of the black professional community in Philadelphia to take a motor trip to Fortescue, New Jersey. Captain Chubb, the motor club's secretary, explained: "The gentleman . . . decided that a real day of outdoor sport, consisting of motoring and fishing would add years to their lives and bring pleasant thoughts to their minds, for some time after."[24]

Advertisements aimed at black motorists told readers that the automobile could bring them closer to nature. The Hollywood Auto Sales Co., for example, ran an article-length advertisement in the *California Eagle* extolling the pleasures of vacationing in a Ford. "Get out the old fish pole, the lunch basket, your camping outfit and camera—you'll need them now that Summer is here again," stated the writer. "Neither stream, nor field, nor mountain park can hide their beauty from you—it's all yours with a Ford." The advertisement detailed the economic merits of the Ford, reminding potential customers of the car's dependability, sturdiness, low cost, easy upkeep, and high resale value.[25] This may have been the standard advertising copy sent to and distributed by all Ford dealers, but the independence and economy of auto touring reappeared in other articles on touring aimed at African-Americans.

Most important, the automobile offered African-Americans escape from the Jim Crow segregation they encountered trains and street cars.[26] In 1930, the writer George Schuyler observed that the main reason African-Americans bought automobiles was to avoid substandard segregated transportation. "All Negroes who can do so purchase an automobile as soon as possible," he wrote, "in order to be free of discomfort, discrimination, segregation and insult."[27] He noted that buying an automobile was a "gesture of manliness and independence." Alfred Edgar Smith, a teacher from Washington, DC, confirmed this desire for independence in 1933. "It's mighty good," he testified, "to be the skipper for a change, and pilot our craft whither and when we will. We feel like Vikings. What if our craft is blunt of nose and limited of power and our sea is macadamized; it's good for the spirit to just give the old railroad Jim Crow the laugh."[28]

The Segregation of a Technological Democracy

In the 1920s, however, black motorists were unable to escape the specter of Jim Crow even on the open road. Beginning with the outdoor recreation movement at the turn of the century, the open road came to symbolize individual freedom. The antithesis of the city, the open road represented a place where American urbanites could escape the degenerative effects of city life and re-create themselves in a community of "like-minded" Americans.[29] Although white travelers constructed the open road as a technological democracy, open to anyone who owned a car, they simultaneously limited access to automobility through a system of discrimination and representation that positioned nonwhites outside the new motor culture.

In their rhetoric, white motor philosophers such as Elon Jessup declared driving and auto camping to be the "most democratic sport in America."[30] For white travelers the social benefits of the new technology were clear: the car gave them a feeling of control over the space of public leisure and entitled them to membership in a national community of middle-class motorists.[31] Observing the regional and class diversity of Americans driving and camping along the open road, Jessup concluded that the car had made pleasure travel available to a broader segment of Americans.

Other white drivers reiterated the idea that the automobile mended class divisions. These travelers characterized the community of the open road as a fraternity and a technological democracy.[32] For example, in 1921 Cornelius Vanderbilt Jr. endorsed the automobile as an agent of democracy. Acknowledging Henry Ford's achievement in lowering the cost of ownership, he wrote that all Americans realized that the "automobile and its manufacturers are helping to solve the . . . social problems of the future." In Vanderbilt's estimation, there were few who had not yet traveled by automobile, and he believed that motor travel brought into "wholesome contact sections of our population that would not ordinarily meet." Drawing on his own experiences, the wealthy capitalist told the readers of *Motor* that he had "brushed shoulders with every type of humanity imaginable" while driving across America. Vanderbilt declared that there was a "poetic justice of the open spaces, for the ways of the road . . . are those of fair play and democracy for all."[33]

But rather than welcoming black motorists, the fraternity of the open road re-inscribed older forms of racial discrimination on the landscape of public leisure. When African-Americans traveled, they encountered not a feeling of national cama-raderie but a contested terrain that was neither officially segregated nor completely welcoming. The growing numbers of white-owned hotels, resorts, and auto camps that grew up alongside the road often did not admit blacks. A writer for *Crisis* sur-veyed the landscape of travel in the United States from the perspective of the black elite: "The white [traveler], for instance, may choose from a thousand different places the one to which he will go with his family. . . . He has but to choose and pay his board. But you, if you are colored, will knock in vain at the farmhouse door for board and lodging. The beautiful, inexpensive, out-of-the-way places are out of your way, indeed."[34]

Alfred Edgar Smith described disappointments for African-Americans on the road. He noted "with good cars growing cheaper . . . there is still a small cloud that stands between us and complete motor-travel freedom. . . . 'Where,' it asks us, 'will you stay tonight?' An innocent enough question for our Nordic friends, of no consequence. But to you and me, what a peace-destroying world of potentiality."[35] After traveling exten-sively in the United States, Smith remarked that racial discrimination against black motorists was not limited to the South: "in spite of unfounded beliefs to the contrary, conditions are practically identical in the mid-west, the South, the so-called North-east, and the South-southwest."[36]

The sociologist Lillian Rhodes added the weight of social research to these personal stories of discrimination. In her pre-Depression study of a middle-class black commu-nity in Pennsylvania, Rhodes surveyed the leisure possibilities for African-Americans. She concluded, "When white Middletown comes back from an extended auto trip, the talk is of the pleasant incidents connected with it. Colored L_____ talks, among other things, of the 'Jim-Crowism' it encounters at rest camps and service stations, and long discussions ensue over the race problem and the probable way out."[37] Another social scientist, Richard La Piere, discovered in 1934 that despite the lack of overt segrega-tion, white auto-camp owners routinely refused black travelers upon arrival.[38] *Negro Motorist Green Book*, a publication written by and for black motorists, conducted a survey of tourist facilities open to black motorists in the 1930s. The majority of respon-dents wrote that they preferred not to advertise to black motorists. An auto camp owner from Montana added that camp managers "believe Negro patrons have a right to fair consideration but they would hesitate to place their names in your directory for fear of finding all touring Negroes near here over-crowding the facilities to the exclu-sion of old customers."[39]

Even the National Parks, the epitome of the open road movement for many white Americans, perpetuated racial discrimination. In the early 1920s, the National Park Service debated whether it could openly exclude blacks. At the sixth National Park Conference, held in Yosemite in 1922, park superintendents "decided that, while colored people could not be openly discriminated against, they should be told that the parks have no facilities for taking care of them." Conference participants reasoned that the Department of the Interior was not at fault if blacks were denied accommodation because "the attitude of the help of the hotels (over whom the park has no jurisdiction) makes it difficult for [black motorists] to go through the park as they cannot be furnished service."[40]

White journalists and travel writers reinforced the exclusion of nonwhites from the growing auto culture by representing them as primitive or nontechnological.[41] For instance, reporting on automobile use among Native Americans for *Illustrated World* in 1922, one journalist observed: "For the most part they abuse their motor cars. They neglect to give the machines proper mechanical attention. . . . Hence the deterioration and depreciation among the red roadsters of the redskins are heavy."[42]

In *Machines as the Measure of Men*, Michael Adas has argued that the portrayal of nonwhites as technologically backward formed the ideological core of white travel narratives.[43] For white Europeans and North Americans differences in material culture, including science and technology, presented a clearer and easier means of identifying cultural superiority than moral measures. Adas argued that for white travelers, explorers, and expansionists evidence of "technological superiority . . . has prompted disdain for African and Asian accomplishments, . . . and legitimized efforts to demonstrate the innate superiority of the white 'race' over black, red, brown, yellow."[44] In the United States, segregation of space as well as representations of blacks as primitive allowed whites to maintain ideas of difference.[45]

Tourist narratives, popular fiction, post cards, and automobile advertising all perpetuated minstrel images of blacks as lazy, boastful, and technologically backward.[46] A common trope of tourist photography was the image of rural nonwhites (African-Americans, Hispanics, Native Americans) standing in front of dilapidated housing, lacking modern technologies such as indoor plumbing or birth control, and baffled by the sight of motor cars.[47] Representing nonwhites as primitive—as noncontributors to modern progress—these images undermined African-Americans' claims to social and technological equality. Furthermore, during the 1920s and the 1930s, white travel writers clung to the image of blacks as reckless drivers and poor mechanics. News coverage of Jack Johnson represented one of the earliest examples, but representations of black drivers as unskilled appeared in a range of cultural sources from fiction to travel

narratives.[48] In a 1920 short story for the *Saturday Evening Post*, humorist Irvin Cobb depicted the misadventures of Red Hoss, a bumbling Sambo figure who took a job as a jitney driver.[49] Behind the wheel of a car, Hoss damaged property and endangered the lives of pedestrians. Inflated with a sense of his own status as a driver, Hoss "began taking sharp turns on two wheels," and in "his first week as a graduate chauffeur he steered his car headlong into a smash-up from which she emerged with a dished front wheel and a permanent Marcel wave in one fender" (figure 1).[50]

Black Responses and Strategies

Although they had high hopes, members of the black elite understood that simply owning an automobile would not free them from segregation or completely revise mis-representations of blacks as reckless or primitive. In 1930, George Schuyler noted that

His Resignation Had
Been Accepted on the
Spot, and the Spot
Was the Bulge of His
Left Jaw

Figure 1
Source: *Saturday Evening Post*, October 9, 1920.

the South was growing accustomed to the notion that blacks owned cars but "there was a time not long ago when in most sections of Dixie a Negro with a well-kept or expensive automobile was suspected of stealing it."[51]

Black leaders used several strategies to overcome discrimination, counter minstrel images of black drivers, and ensure participation in the new motor culture. Middle-class blacks in the 1920s and the 1930s mounted a tripartite campaign for auto citizenship: they produced counter images of black drivers as inventive and respectable; they legally challenged discrimination by auto insurance companies and hotels; and they created separate systems of travel that protected the comfort and safety of black motorists.

In the 1920s, the black press worked hard to counter stereotypes by producing images of the black motorist as technologically skilled and socially responsible. Kevin Gaines studied the black community's use of photography to construct images of racial uplift, "ideal types of bourgeois black manhood and womanhood," and refute "negro-phobic caricatures."[52] Gaines noted that African-Americans sought to "infuse the black image with dignity, and to embody the 'representative' Negro by which the race might more accurately be judged. . . . Anything less than stylized elegance would betray the ideals of race advancement." These images often incorporated attributes of cleanliness, leisure, and literacy.[53] Stories and photographs of black motorists added characteristics of ingenuity and economic success to images of racial uplift.

Eager to demonstrate black mechanical expertise in a culture that valued automotive know-how, black journalists reported the activities of black automobile drivers, inventors, and entrepreneurs. Black newspapers appropriated the dominant discourse that portrayed test drivers as the epitome of masculinity, expressed both physically and technologically.[54] For instance, when A. L. Headen drove his new car from Chicago to Kansas City in July 1922, the *Chicago Defender* lauded both the superior design of the car and Headen's technological expertise and physical prowess. The grueling trip tested the endurance of the driver as much as it did the car. The paper praised Headen for driving with the "tenacity of a bulldog," and with possessing the same steely strength as his car. Headen was not only a skilled driver but also a successful inventor, and the *Defender* called his car design "a masterpiece of automobile engineering." As an inventor, Headen ostensibly had the technological credentials to claim full access to the privileges of automobility. According to social psychologist and patent examiner Joseph Rossman, who challenged existing notions of racial inferiority in the late 1920s, any race that produced inventors could not be inferior, or at least had the potential to contribute to the progress of civilization.[55] Technological skill could blur the lines of racial and class difference and underpin arguments for social equality.

The black press also recorded the contributions of African-Americans to motoring as auto mechanics, inventors, entrepreneurs, and driving instructors. Benjamin Thomas, an inventor and driving instructor, observed in 1937, "In New York City alone, one third of the Mechanical work is being done by Colored men, and the same that applied to New York, applies to all other cities and towns through the country." He continued, "The Negro owns and controls a large percentage of the taxicab business throughout the country, and the garage and repair shop business is held in the same proportion."[56] The black community had its share of automobile experts in the 1920s and the 1930s. However, inventors and mechanics such as Headen and Thomas were recognized neither by the white automotive press nor by the mainstream motor travel authorities.

Automobiles offered a springboard into the middle class for African-Americans both economically and culturally. Chauffeuring and, later, jitney services became a way for black mechanics and drivers to own their own businesses and achieve status comparable to, if not above, that of Pullman porters. Social commentators such as Charles Johnson argued in the 1940s that in the early days of automobility, "negroes were expected to operate automobiles for whites, but not to own them."[57] Yet, historically, a small group of black inventors and entrepreneurs not only owned cars in the 1910s and the 1920s but took advantage of their automotive skill to run driving schools for chauffeurs and even start their own automotive companies.[58] Lee Pollard, for instance, opened a garage and automobile school in New York City in 1901, which blossomed into a training institute for black chauffeurs in the 1910s. Pollard, who published a manual for chauffeurs in 1914 entitled "How to Run an Automobile," made a respectable income by teaching others how to drive. He also gained renown in New York's black community as an inventor of automobile accessories, including a convertible enclosure, and later as a builder of airplanes and owner of an African import business.[59]

The *California Eagle* announced the rise of black-owned automobile enterprises as a step forward for African-Americans.[60] The automobile had become, for this Southern California community, both a business opportunity and a personal necessity. The newspaper observed in March 1924: "Time was when automobiles were a curiosity in so far as owning one by any number of the group was concerned, but time has wrought many changes and the automobile is a necessity to every one."[61] *Eagle* editor, E. L. Dorsey was so moved by the increasing numbers of automobiles among his readers that he conducted an informal survey of black car owners in 1924. Although the sample was limited to a small group of his friends and colleagues, the editor insisted that his data would not "deviate in the least" from a national average.[62] Dorsey's findings antici-

pated the results of later national surveys—namely, that the majority of black motorists owned other property, such as a house or land, and that a majority of black car owners depended on the car for their livelihood. Unfortunately, Dorsey's survey was a unique endeavor in the 1920s; automobile manufacturers and the U.S. Department of Commerce did not recognize or study black consumers until the Depression. [63]

Dorsey used his data to counter cultural prejudices against black drivers. He refuted the notion that "the automobile is a luxury overly indulged in by many of our group," and argued that "the average person was entitled to enjoy the distinction of owning an auto, if he was able to own a home." Dorsey argued that his readers ascribed to the same values as the white middle class, including property ownership and prudent economic practice. He also challenged the idea that middle-class black motorists were reckless drivers, noting that "it is erroneous to think that the group is a bunch of joy riders, so let us before accusing our auto owners of such investigate and we may find that after all the average auto is owned by persons fully able to financially carry the burden rather than by the worthless element." [64]

Automobile ownership provided material evidence that one segment of the black community had achieved a middle-class life style, while, at the same time, reinforcing class and color divisions among African-Americans. [65] The *California Eagle* published news of the most elite strata of automobile owners in its "Automotive Section." In the early 1920s, the *Eagle's* "Exhaust" column functioned as a social register of automobile owners by recording who bought new cars and who took automobile vacations. Listing the occupations and leisure activities of automobile owners, the paper wove the automobile into portraits of black success. For instance, in December 1923 readers learned that "S. B. W. Mayo, besides steering the destinies of the Citizen's Investment Company also steers a Studebaker; Dr. Ruth Temple, besides being one of our leading medical lights is a believer and leader in automobiles; and Wm. Pickens and Bob Davis, who move folks for a livelihood, have two sport Durants, and believe us, when we say they sport some." [66]

Photographs of car owners often accompanied the social bylines in the Exhaust column as evidence of black participation in a booming automobile market. "Prosperity in a marked degree is on parade," declared Dorsey in December 1923. As evidence for his point, Dorsey offered a photo of Miss Hill and her Reo Sport Phaeton (figure 2). [67] Another picture featured Joseph Marable and John Williamson posing with their sport Kissell in front of the *Eagle* office in the summer of 1924. Embarking on a cross-country tour to Boston, Marable and Williamson wore fashionable suits, ties, and hats (figure 3). [68] Like studio portraits of sober families in bourgeois domestic settings, these snapshots represented the black driver as respectable and prosperous; the subjects were

Figure 2
Source: *California Eagle*, December 21, 1925.

well-dressed and the cars were in pristine condition. Such photos presented a middle-class ideal of material success, leisure, and responsible car ownership.

Alice Dunbar-Nelson, writer, advocate of racial uplift, and former wife of poet Paul Dunbar, offered a more complex view of car ownership among her peers in the 1920s than the one presented in the *California Eagle*.[69] For Dunbar-Nelson owning an automobile was part of maintaining a middle-class life, similar to gaining a college education or owning a house. Running a struggling newspaper, the Wilmington *Advocate*, Dunbar-Nelson and her husband were chronically short on money. She wrote in a moment of despair that material possession and position in the community rather than income marked her as middle class: "I have no money, and things are far from going well with us. . . . [Yet] to all intents and purposes I prosper, so far as the world can see." A creative negotiator, Dunbar-Nelson shared her car and its expenses with an employee of the *Advocate* and later shared a car with her niece. This strategy proved difficult, interfering with the ideal of personal mobility. On the other hand, for Dunbar-Nelson car

Figure 3
Source: *California Eagle*, June 27, 1924

ownership was well worth compromises, scrimping, and even the trouble of passing the State driver's exam.[70] Like other black activists in the 1920s and the 1930s, she used the car to avoid segregated trains while on speaking tours and to take vacations to the mountains or the beach. For Dunbar-Nelson and her peers, by the 1920s the car had become a signifier of class status, a instrument of pleasure, and an item of necessity.

As the car became part of the everyday life of more African-Americans, the black elite launched a series of legal challenges to discrimination on the road. With the help of the NAACP and the Urban League, black motorists attempted to end the discriminatory policies of auto insurance companies and hotels. Representations of blacks as reckless drivers had had material consequences for African-American motorist, who had great difficulty acquiring automobile insurance in the 1920s and the 1930s.

Wendell Sayers, a real estate broker in Seattle, wrote to Thurgood Marshall, special counsel for the NAACP, in the fall of 1938 seeking a list of insurance companies that would accept blacks. "As a broker," Sayers wrote, "I have applications for automobile liability insurance and property damage insurance on an automobile owned and operated by Negroes and Orientals. In this section of the country, I am advised that insurance companies are refusing this risk because of prejudice, many of them stating frankly why they refuse the risk."[71] Sayers added that he was black and that he considered it his duty to his race to help his clients find insurance companies. The NAACP heard the same story from several states,[72] and both the NAACP and the Urban League eventually initiated legal challenges to insurance companies covering automobile liability.[73] Marshall and others noted that the denial of liability insurance put an unnecessary burden on blacks by forcing them to prove their financial security, pay higher premiums, and by denying them a driver's license if they could not acquire insurance.[74]

Various members of the black community also brought suit against discriminatory policies at hotels in the 1930s. One lawsuit initiated by a Washington physician, Dr. B. Hurst, against a New York City hotel prompted a broader campaign against discrimination by hotels. Motoring from Montreal to New York City, Hurst's experiences fit a pattern all too familiar to blacks. Although the family made advance reservations at the Prince George Hotel in New York City, when the group arrived at midnight, the hotel night clerk refused them service. Saying that he never received a room reservation, the clerk turned Hurst and his party out onto the street. In a detailed letter to Walter White, secretary of the NAACP, Hurst described the nightmare of searching the dark city for rooms. After being refused at four different hotels he "elected to try [his] fortunes in the 'Y's' of Harlem."[75]

Hurst's experience gave the NAACP the perfect upper middle-class subject to challenge racial discrimination at hotels. Walter White identified Hurst as an "excellent test case" because "Dr. Hurst is a distinguished physician. . . . He, his mother, and Dr. Taylor are fine persons, cultured and intelligent."[76] The injustice of the case would be based as much on class status as on race. Hays suggested that the NAACP file suit against all the hotels involved, using the New York Civil Rights law. Hays and his clients won their suit plus damages in the amount of $400. Most importantly, Hurst's case touched off other similar litigation and eventually led to a petition distributed by the Federal Council of Churches in 1936 that called for white hotel owners to "remove restrictions upon the acceptance of colored people as guests. . . [and] the acceptance of colored and mixed groups for dinners. . . and other affairs customarily held in hotels."[77] Despite these early efforts, hotel discrimination persisted throughout the 1930s; landmark legislation did not come until 1964, when the *Heart of Atlanta Motel vs. United*

States case made discrimination in the operation of public accommodations illegal under free-commerce laws.[78]

Facing continued discrimination on the open road, African-Americans created separate systems of travel, including resorts, hotels, and travel agents. Fueled by a newly motorized black middle class, resorts such as those at Idlewild, Michigan, and Elsinore, California, flourished in the 1920s.[79] Although the Idlewild was founded by white developers, members of the black community, such as W. E. B. Du Bois, endorsed it as a haven of summer pleasure for African-Americans.[80] In a 1921 review of possible tourist destinations for black travelers, Du Bois told the readers of *Crisis*: "I know the cost and prejudice and intriguing ugliness of Atlantic City. I have tasted the lovely beauty of the beach at Sea Isle and sat in the pretty dining room at Dale's, Cape May. . . . Beside Idlewild they are nothing. Not for one moment in fine joy of life, absolute freedom from the desperate cruelty of the color line . . . not for one little minute can they rival or catch the bounding pulse of Idlewild."[81]

Less well-known patrons of Idlewild encouraged members of the black middle class in Chicago to invest in the resort. In an advertising booklet published by the Idlewild Resort Company, various African-American doctors, dentists, and ministers praised Idlewild as a motor destination. "I had the opportunity last summer of motoring from Chicago to Idlewild," wrote Dr. Hale Parker, a dental surgeon from Chicago in 1919. "The trip in itself was a pleasure as the roads were extremely good."[82] Others added that they enjoyed living like gypsies and motoring within the confines of the resort. These pleasure seekers fulfilled Du Bois's mandate to "develop, beautify, and govern" this space.[83] For those within a day's drive of Idlewild, the resort offered the best of the outdoor movement because it eliminated the problems of searching for a place to stay.

At the same time, there were some efforts to create protective associations for black motorists who wished to undertake trips of more than a day. For instance, black auto racing enthusiasts in Chicago formed the AAAA, the Afro-American Automobile Association, in 1924. Leaders in the black automobile community of Chicago included Headen, the automobile inventor, auto driver, and head of his own automobile manufacturing company, who wanted to create an organization that would promote black automobile racing.[84] Recognizing the particular needs of black motorists in general, the organizers enlarged the mission of the AAAA to include assistance to travelers on a national scale. The *Chicago Defender* reported that the organization hoped to enlist "the aid and support of all Race auto owners and drivers throughout the country. The ultimate aim is to establish branch offices with members in all principal cities and business relations that will be of material benefit to the members when traveling."[85]

Although the founders envisioned an organization that would serve black motorists across the nation, the AAAA faded from the pages of the black press after the initial announcement of its founding.

In 1936, Victor Green, a black travel agent in New York City, answered the call for a national guidebook of hotels and other motor services that were either black-owned or catered to black tourists. He explained, "it has been our idea to give the Negro traveler information that will keep him from running into difficulties, embarrassments and to make his trip more enjoyable."[86] The *Negro Motorist's Green Book* began as a local publication in 1936, but the desire for copies was so great that the booklet was sold nationally in 1937. Widening the scope of information provided, Charles McDowell, collaborator on Negro Affairs for the United States Travel Bureau (also created in 1937), contributed national information to the publication.[87] The publication offered advice on automotive upkeep, but its primary duty was to survey the difficult terrain of the open road and provide accurate information on hotels and other accommodations that would accept black motorists. And at the end of the day, it helped answer the question, "Where will you spend the night?" In this way, the *Negro Motorist's Green Book* was a pioneering effort to make automobility possible for middle-class blacks. The travel guide, along with the auto-related services it contained, helped create a safe, if separate, road for African-American motorists in the period just before World War II, and on into the years after 1945.[88]

Conclusion

More than 80 years after Jack Johnson's repeated arrests, race continues to structure automobility in the United States. Racial profiling, or the crime of "driving while black," continues to cast black drivers as potential threats to public safety. Recent surveys in several American cities have shown that African-Americans are stopped by police much more often than other drivers.[89] In 1999 the magazine *Emerge* offered this observation: "Members of Congress, athletes and entertainers, even attorneys, all have related tales of what is commonly know as DWB."[90] Limiting mobility and continuing racial stereotypes, the issue of "driving while black" has become a rallying point for African-American leaders who continue to struggle for the individual autonomy promised by the car.[91]

This study of black motorists in the early twentieth century describes the world of automobility, and of the open road, as racially contested terrain. It also provides historical context for understanding the appropriation of the automobile in that period by African-Americans to challenge discrimination in spatial and technological terms.

Facing racial prejudice and hostile misrepresentation, black political leaders and news-paper editors used the automobile as an icon of the technological skill and the economic power of the African-American middle class, and to refute white stereotypes of black drivers. As they purchased, drove, and tinkered with cars in the 1920s, African-Americans inserted themselves into a larger national discourse of technological progress—claiming ingenuity, bourgeois responsibility, and, ultimately, equal citizenship in the technological democracy of the open road.

Notes

1. "Jack Johnson Wants a Race," *New York Times*, July 31, 1910. For another historical account of the race, see Michael L. Berger, "The Great White Hope on Wheels," in *The Automobile and American Culture*, ed. D. Lewis and L. Goldstein (University of Michigan Press, 1991).

2. Jack Johnson, *Jack Johnson Is a Dandy: An Autobiography* (Chelsea House, 1969), p. 152; Al-Tony Gilmore, *Bad Nigger! The National Impact of Jack Johnson* (Kennikat, 1975), p. 21.

3. "Oldfield Here to Race Johnson," *New York Times*, October 10, 1910, p. 10, col. 6; "Barney Oldfield Accepts Negro Fighter's Challenge to Compete Here," *New York Times*, October 5, 1910.

4. Gail Bederman, *Manliness and Civilization: A Cultural History of Gender and Race in the United States, 1880–1917* (University of Chicago Press, 1995), pp. 1–5; Randy Roberts, *Papa Jack: Jack Johnson and the Era of White Hopes* (Free Press, 1983), pp. 68–69, 85–107. On the history of boxing and its embodiment of male virility in the 1890s, see Elliot Gorn, *The Manly Art: Bare-Knuckle Prize Fighting in America* (Cornell University Press, 1986), pp. 194–206.

5. On the relationship between race, manliness and civilization, see Bederman, *Manliness and Civilization*, pp. 23–31. On the central role of technology in twentieth-century conceptions of American civilization and progress, see Carroll Pursell, *The Machine in America: A Social History of Technology* (Johns Hopkins University Press, 1995), pp. 229–250. For an earlier and broader history of the links between civilization, white racial dominance, and technology, see Michael Adas, *Machines as the Measure of Men: Science, Technology and Ideologies of Western Dominance* (Cornell University Press, 1989), p. 15.

6. "Jack Johnson Answers Auto Speeding Charge in Muncie Court," *New York Times*, April 2, 1910; "Johnson a 'Good Family Man,'" *New York Times*, April 6, 1910; "Johnson Defies Police," June 23, 1910; "Johnson Fails to Appear in Court," *New York Times*, June 24, 1910 and July 3, 1910; "'Jack' Johnson Arrested, Accused of Reckless Driving," *New York Times*, July 20, 1910; "Court Warns Johnson," *New York Times*, July 21; "Jack Johnson Fined for Speeding in Cleveland," *New York Times*, August 19, 1910.

7. See James Flink, *The Automobile Age* (MIT Press, 1988), p. 132; Kenneth Jackson, *Crabgrass Frontier: The Suburbanization of the United States* (Oxford University Press, 1985), pp. 157–158.

8. For an history of the relationship between a rising black middle-class, the growth of segregation, and the construction of whiteness as a spatial issue, see Grace Elizabeth Hale, *Making Whiteness: The Culture of Segregation in the South, 1890–1940* (Pantheon, 1998).

9. On the replacement of culture by technology, see Pursell, *The Machine In America*, pp. 229–230.

10. David Langum, *Crossing Over the Line*: *Legislating Morality and the Mann Act* (University of Chicago Press, 1994), pp. 179–186, 247.

11. Kevin Gaines, *Uplifting the Race*: *Black Leadership, Politics, and Culture in the Twentieth Century* (University of North Carolina Press, 1996), pp. 3–4.

12. For historians who have considered both the design and consumption of new technologies as political, see Steven Lubar, "Machine Politics: The Political Construction of Technological Artifacts," in *History From Things*: *Essays on Material Culture*, ed. S. Lubar and W. Kingery (Smithsonian Institution Press, 1993); Langdon Winner, "Do Artifacts Have Politics?" in Winner, *The Whale and the Reactor: A Search for Limits in an Age of High Technology* (University of Chicago Press, 1986); Wiebe E. Bijker, Thomas P. Hughes, and Trevor Pinch, "The Social Construction of Facts and Artifacts, " in *The Social Construction of Technological Systems*, ed. W. Bijker et al. (MIT Press, 1987). On the marketing and consumption to new technologies, see Carolyn Marvin, *When Old Technologies Were New: Thinking About Electric Communication in the Late Nineteenth Century* (Oxford University Press, 1988), pp. 4–5; Susan Douglas, *Inventing American Broadcasting, 1899–1920* (Johns Hopkins University Press, 1987). For models on using material culture to explore social class, see Stuart Blumin, *The Emergence of the Middle-Class: Social Experience in the American City, 1760–1900* (Cambridge University Press, 1989); Lizabeth Cohen, "Embellishing a Life of Labor: An Interpretation of Material Culture of American Working Class Homes, 1885–1915," *Journal of American History*, winter 1980: 752–775.

13. For explorations of gender and the history of technology that politicize the consumption of new machines, see Sherrie Inness, "On the Road and in the Air: Gender and Technology in Girls' Automobile and Airplane Serials, 1909–1932," *Journal of Popular Culture* 30 (1996), fall: 37–46; Virginia Scharff, *Taking the Wheel: Women and the Coming of the Motor Age* (Free Press, 1991) and Ruth Schwartz Cowan, *More Work For Mother: The Ironies of Household Technology from the Open Hearth to the Microwave* (Basic Books, 1983). On gender as methodological approach that has expanded the history of technology, see Carroll Pursell, "Seeing the Invisible: New Perceptions in the History of Technology," *Icon* 1 (1995): 9–15; Lana F. Rankow, "Gendered Technology, Gendered Practice," *Cultural Studies in Mass Communication* 5 (1988): 57–70; Nina Lerman, Arwen Palmer Mohun, and Ruth Oldenziel, "Versatile Tools: Gender Analysis and the History of Technology," *Technology and Culture* 38 (1997), January: 1–9.

14. On the idea that the black and white middle classes used the auto in similar ways, see Blaine Brownell, "A Symbol of Modernity: Attitudes Toward the Automobile in Southern Cities in the 1920s," *American Quarterly*, March 1972, p. 33.

15. Bart Landry, *The New Black Middle Class* (University of California Press, 1987), p. 21. See also E. Franklin Frazier, *Black Bourgeoisie: The Rise of a New Middle Class* (Free Press, 1957), p. 45.

16. For a description of the black middle class in 1920, see Landry, *The New Black Middle Class*, p. 43.

17. On the class expectations of the New Negro, see Kevin Gaines, *Uplifting the Race: Black Leadership, Politics, and Culture in the Twentieth Century* (University of North Carolina Press, 1996).

18. Robert Russa Moton, *What the Negro Thinks* (Doubleday, 1929), pp. 37, 42.

19. On rising wages and automobile ownership among black tenant farmers, see Arthur E. Barbeau and Florette Henri, *The Unknown Soldiers: Black American Troops in World War I* (Temple University Press, 1974), p. 10.

20. Melnotte C. Wade, "That Auto of Mine," *Champion*, January 1917, p. 258.

21. Phrase from Wilbur Hall, "A Free Car and the Open Road: Auto-Vacationing in Three States Along the Pacific," *Sunset*, July 1917, p. 22.

22. Gaines, *Uplifting the Race*, p. 70.

23. "Where Will You Spend Your Vacation?" *California Eagle*, June 13, 1924; June 20, 1924.

24. "The Events of Yesterday and the Happenings of Today," May 29, 1924, in *The Journal of C. Chubb* (Library Company of Philadelphia).

25. "To Make Pleasure for the Whole Family," *California Eagle*, August 1, 1924.

26. William Pickens, "Jim Crow in Texas," *The Nation*, August 15, 1923, p. 156.

27. George S. Schuyler, "Traveling Jim Crow," *American Mercury*, August 1930, p. 432.

28. Alfred Edgar Smith, "Through the Windshield," *Opportunity*, May 1933, p. 142. For Smith's occupation, see *Polk City Directory*, Washington, DC, 1910–1935.

29. Bruce Calvert, *Thirty Years of the Open Road with Bruce Calvert* (Greenberg, 1941), pp. 2, 64–65; "Bruce Calvert, 73, Publisher, Writer," *New York Times*, June 1, 1940. See also Bok, "Summers of Our Discontent," 16; Dillon Wallace, "Open Spaces on the Map," *Outing*, April 1910, p. 570; Lucy E. Abel, *The Open Road and Other Poems* (Gorham, 1916), p. 7; "Call of the Open Road," *Country Life*, June 1923, p. 112; F. G. Jopp, "Call of the Open Road," *Illustrated World*, July 1922: 742–746; August 1922: 902–905; Henry Ford, "The Modern City: A Pestiferous Growth," in *Being A Selection from Mr. Ford's Page in the Dearborn Independent* (Dearborn, 1926), pp. 156, 158. Owen Tweedy, "The Fellowship of the Open Road," *Atlantic*, September 1932, p. 321.

30. Elon Fessup, "The Flight of the Tin Can Tourists," *Outlook*, May 25, 1921, n.p.

31. Hall, "A Free Car and the Open Road," pp. 22–26.

32. "Automobiling Has Not Bred Anarchy," *American Motorist*, November 1910, p. 500; Frederic F. Van De Water, "Discovering America in a Flivver," *Ladies Home Journal*, May 1926, p. 28; Nina Wilcox Putnam, *West Broadway* (Dornan, 1921), pp. 83–84.

33. Cornelius Vanderbilt Jr., "The Democracy of the Motor Car," *Motor*, December 1921: 21–22.

34. "Vacation Days," *The Crisis*, August 1912, p. 186.

35. Smith, "Through the Windshield," 142.

36. Ibid., pp. 142, 143.

37. Lillian Rhodes, "One of the Groups *Middletown* Left Out," *Opportunity*, March 1933, p. 76.

38. Richard T. La Piere, "Attitudes vs. Actions," *Social Forces* 13 (1934), December: 231. La Piere's test subjects were Chinese Americans, but the initial study was inspired by problems of black travelers.

39. Victor Green, *The Negro Motorist Greenbook* (Victor Green, 1938), p. 4.

40. National Park Service, *Summary of Decisions Reached at the Various Conferences of National Park Superintendents 1911–1935* (Department of the Interior, Washington, 1935) Record Group 79, Central Classified Files, National Park Service, 1907–1935, National Archives II, College Park, Maryland.

41. Winifred Dixon, *Westward Hoboes* (Scribners, 1921), pp. 164, 180; Bedell, *Modern Gypsies*, p. 90; A. C. Laut, "Rediscovering America," *Technical World*, December 1914, pp. 488, 492.

42. George H. Dacy, "Redskins and Red Roadsters," *Illustrated World*, August 1922, pp. 862, 864. On Native American ingenuity as an exception, see "A Full-Blooded Indian Ford," *Ford Owner and Dealer*, October 1924, p. 54.

43. Adas, *Machines as the Measure of Men*, pp. 154, 159.

44. Ibid., p. 15.

45. Hale, *Making Whiteness*, p. 130.

46. For an examination of the image of the Southern black as poor, rural, and nontechnological as a marketing strategy for the New South, see Hale, *Making Whiteness*, p. 44. On racial inequality and tourism in the United States, see Robert Rydell, *All the World's a Fair: Visions of Empire at American International Expositions, 1876–1916* (University of Chicago Press, 1984). For an explanation of construction of racial difference and the tourist gaze, see John Urry, *The Tourist Gaze: Leisure and Travel in Contemporary Societies* (Sage, 1990); James Clifford, *The Predicament of Culture: Twentieth-Century Ethnography, Literature, and Art* (Harvard University Press, 1988).

47. For tourist representation of Southern blacks as primitive, see Mary Bedell, *Modern Gypsies*, p. 379; "C.T.'s Happy South Scenes," postcard, Victor A. Blenkle Collection, box 26, Archive Center, National Museum of American History; "Race Suicide Down South," postcard, Hoffman Boaz Collection, Archives Center, NMAH; Jay Norwood Darling, *The Cruise of the Bouncing Betsy: A Trailer Travelogue* (Frederick Stokes, 1937), p. 50.

48. Beatrice Massey, *It Might Have Been Worse: A Motor Trip from Coast to Coast* (Harr Wagner, 1920), p. 22; Andrew Wilson, *The Gay Gazel: An Adventure in Auto Biography* (privately printed by the author, 1926), pp. 15, 17.

49. On Cobb's life and work, see Wayne Chatterton, *Irvin S. Cobb* (Twayne, 1986).

50. Irvin Cobb, "A Short Natural History," *Saturday Evening Post*, October 9, 1920, p. 5. On heightened racial stereotypes in popular culture in the 1920s, see Melvin Patrick Ely, *The Adventures of Amos 'n' Andy: A Social History of an American Phenomenon* (Free Press, 1991), p. 25.

51. Cobb, "A Short Natural History," p. 6.

52. Gaines, *Uplifting the Race*, p. 68.

53. Ibid., p. 69.

54. Nolan, *Barney Oldfield*, p. 113.

55. Joseph Rossman, "The Negro Inventor," *Journal of the Patent Office Society* 11 (1928), pp. 552, 553.

56. Benjamin Thomas, "The Automobile and What It Has Done for the Negro," in *The Negro Motorist Greenbook* (Victor Green, 1938), p. 11; "Conduct Automobile Schools in New York:

Several of the Best Garages and Schools are Owned by Negroes," *New York Age*, February 13, 1913. See also "Ben Thomas Invents an Auto Enclosure," *New York Age*, September, 16, 1915.

57. Charles Johnson, *Patterns of Negro Segregation* (Harper, 1940), p. 124.

58. Jill D. Snider, Flying to Freedom: African American Visions of Aviation, 1910–1927, Ph.D. dissertation, University of North Carolina, 1995, pp. 24, 32.

59. Ibid., p. 25; "The Easy Way: How to Run an Automobile," *New York Age*, July 2, 1914.

60. "Auto Enterprises among Race Grow," *California Eagle*, August 15, 1924; "Red Top Taxi Company," *California Eagle*, February 14, 1924.

61. "Shall We Have a Show?" *California Eagle*, March 21, 1924. See also "Automobile Industry Has A Most Prosperous Year," *California Eagle*, December 21, 1923.

62. "The Owner of the Auto?" *California Eagle*, July 25, 1924.

63. These general surveys were generated in the 1930s as an attempt to understand an overlooked population and open new markets during the Depression. On automobile ownership, during the 1930s, see Richard Sterner, *The Negro's Share: A Study of Income, Consumption, Housing, and Public Assistance* (Harper, 1943), pp. 144, 147. On income, recreation, and the purchasing power of blacks in the interwar period, see Paul K. Edwards, *The Southern Urban Negro as a Consumer* (Prentice-Hall, 1932); Eugene Kinckle Jones, "Purchasing Power of Negroes In the U.S. Estimated at Two Billion Dollars," *Domestic Commerce*, (U.S. Department of Commerce, 1935), p. 3468; "The Negro Market," *Opportunity*, February 1935, p. 38. For a bibliography of articles on black-owned business in the interwar period, see U.S. Department of Commerce, *The Negro in Business* (Bureau of Foreign and Domestic Commerce, 1935), pp. 1–3.

64. "Owner of the Auto?" *California Eagle*.

65. Gaines, *Uplifting the Race*, pp. 2–4.

66. "Exhaust," *California Eagle*, December 21, 1923. For more examples, see "Exhaust," *California Eagle*, February 17, 1923; "Exhaust," *California Eagle*, June 20, 1924; "Exhaust," *California Eagle*, August 15, 1924.

67. California *Eagle*, December 21, 1923.

68. "Leave on Auto Trip to Boston," *California Eagle*, June 27, 1924.

69. Gloria T. Hull, ed., *Give Us Each Day: The Diary of Alice Dunbar-Nelson* (Norton, 1884).

70. Ibid., pp. 240, 323, 326, 375–380.

71. Wendell P. Sayers to Thurgood Marshall, September 27, 1938, Discrimination File, Auto Insurance, NAACP Collection, Library of Congress.

72. Dick Reynolds to the NAACP, January 7, 1939; Dick Reynolds to NAACP, May 28, 1939; Thurgood Marshall to Dick Reynolds, April 3, 1939, Discrimination File, Auto Insurance, NAACP Collection.

73. C. W. Fairchild, Association of Casualty and Surety Executives, to Honorable Louis H. Pink, Superintendent of Insurance, State of New York, May 25, 1939; Thurgood Marshall, New York, to Dick Reynolds, April 3, 1939; Thurgood Marshall to Clarence Mitchell, St. Paul Urban League, April 27, 1939; Clarence Mitchell, St. Paul, to Thurgood Marshall, May 2, 1939, Discrimination File, Auto Insurance, NAACP Collection.

74. Dick Reynolds to the NAACP, New York, April 26, 1939; Carl Erickson to Clyde B. Helm, Secretary Treasurer, Insurance Federation of Minnesota, April 19, 1939; Thurgood Marshall to

Honorable Charles Poletti, Lieutenant Governor of New York, April 6, 1939; Louis Pink, Superintendent, State of New York Insurance Department to Lieutenant Gov. Charles Poletti, April 20, 1939, Discrimination File, Auto Insurance, NAACP Collection.

75. Price Hurst to Walter White, New York, September 8, 1932, Discrimination files: Hotels, Prince George Hotel, 1932–33, NAACP Collection.

76. Walter White, New York, to Arthur Garfield Hays, Esq., September 23, 1932; Walter White, New York, to Arthur Spingarn, September 20, 1932, Discrimination file, Hotels, Prince George, 1932–33, NAACP Collection.

77. Memo to the Members of the Committee on Hotel Arrangements, April 24, 1936, and Walter White, New York, to Elmer A Carter, October 25, 1934, Discrimination file: Stevens Hotel, 1929–1930, NAACP Collection.

78. For a synopsis of that case, see *West's Encyclopedia of American Law*, vol. 3 (West, 1998), p. 96.

79. On Elsinore, see "Where Will You Spend Your Vacation?," *California Eagle*, June 27, 1924; "Elsinore Makes Ready to Entertain Auto Tourists," *California Eagle*, February 14, 1924.

80. Benjamin C. Wilson, "Idlewild: A Black Eden in Michigan," *Michigan History* 65 (1981), September-October: 33–37.

81. W. E. B. Du Bois, "Hopkinsville, Chicago and Idlewild," *The Crisis*, August 1921, p. 160.

82. *Idlewild Resort Company* (Chicago Historical Society, c. 1920).

83. Du Bois, "Hopkinsville, Chicago and Idlewild," p. 160.

84. On L. A. Headen and his Motor Company, see "Headen Breaks Record Running to Kansas City," *Chicago Defender*, July 8, 1922; "Headen Motor Company," *Chicago Defender*, August 26, 1922; "Headen Motor Company Elects New Officers," *Chicago Defender*, January 13, 1923; "Headen Motor Company Elects Officers," *Chicago Defender*, January 19, 1924.

85. "Form Automobile Association," *Chicago Defender* (weekly edition), September 27, 1924.

86. Victor Green, *The Negro Motorist Green Book* (Victor Green, 1948), p. 1.

87. U.S. Department of the Interior, *The United States Tourist Bureau* (Government Printing Office, 1937), p. 5. See also U.S. Tourist Bureau, *Directory of Negro Hotels and Guest Houses* (Government Printing Office, 1941).

88. On the continuing difficulties of traveling in the postwar period, see Kris Lackey, "The Nigger Window: Black Highways and the Impossibility of Nostalgia," in *Road Frames*, ed. K. Lackey (University of Nebraska Press, 1997); Ray Springle, *In the Land of Jim Crow* (Simon & Schuster, 1949), p. 29. On black travel in the 1950s, see H. Sutton, "Negro Vacations: A Billion Dollar Business," *Negro Digest*, July 1950: 25–27; "Vacation Guide," *Ebony*, June 1954, p. 83; L. A. Nash, "Ten Year Milestone for Travelers," *The Crisis*, October 1955: 459–463; "Europe on a Shoestring," *Our World*, December 1954: 28–33; "Trailer Park Landlord,'" *Ebony*, August 1956: 107–110; "Travel Bureau Hits Jack Pot: Philadelphian Operates the Largest Negro Owned Travel Agency," *Color*, May 1956, p. 8. See also Cathey Mac Gregor, ed., *Travel Guide* (privately printed, 1947), sold with the slogan: "Vacation and Recreation without Humiliation."

89. For stories of Driving While Black in local newspapers from 1997 to 1999, see Lee Daniels, "Is DWB (Driving While Black) America's Newest Crime?" *Chicago Defender*, July 9, 1997; "Driving While Black," *Detroit News*, April 26, 1998; "Presumed Guilty," *Houston Chronicle*, August 20, 1998; Marcus Amick, "Motorist's Stop Highlights Issue of 'Driving While Black,'"

Michigan Chronicle, July 1, 1998; James T. Campbell, "Driving While Black Still Gets You Stopped," *Houston Chronicle*, September 14, 1998; Elizabeth Mehren, "Judge Cites Man's Record of 'Driving While Black,'" *Los Angeles Times*, December 17, 1998; David Cole, "Driving While Black: Curbing Race-Based Traffic Stops," *Washington Post*, December 28, 1998; David Dante Troutt, "Hidden Price for 'Driving While Black,'" *Houston Chronicle*, January 31, 1999; "Driving While Black," *San Francisco Chronicle*, March 26, 1999. For legal analysis of Driving While Black and racial profiling as an violation of the Fourth Amendment, see David A. Harris, "Driving While Black: Unequal Protection under the Law," *Chicago Tribune*, March 11, 1997; David A. Harris, "'Driving While Black' and All Other Traffic Offenses: The Supreme Court and Pretextual Traffic Stops," *Journal of Criminal Law and Criminology* 87 (1997), winter: 544–582.

90. "Decriminalizing 'Driving While Black,'" *Emerge*, January 31, 1999, p. 26.

91. Ali Rahman, "It's a Case of Being Shot While DWB—Driving While Black!" *New York Beacon*, May 20, 1998.

The Matter of Race in Histories of American Technology

Rebecca Herzig

In the wake of more than 30 years of feminist studies of artifacts, scholars are now beginning to acknowledge the mutual shaping of gender and technology. We no longer simply ask how women or men use artifacts such as telephones or space shuttles, or how artifacts such as telephones or space shuttles might differentially impact women or men. Rather, we have learned to examine how technologies fabricate the very shape and substance of sexed bodies, even as the persistent sexing of bodies transforms the built world. Consider, for instance, anthropologist David Hess's fine account of the redesign of the Ford Probe automobile in the early 1990s.[1] When Ford executive Mimi Vandermolen was assigned control over the project, she consulted a number of U.S. professional women for their suggestions and recommendations. Subsequent amendments to the new car included gas pedals set at a right angle to aid drivers wearing high heels, contoured knobs for drivers with long fingernails, a new molding on the seats that would not snag dresses, and a lowered cowl (the place where the windshield meets the engine compartment), since "tests showed that women generally liked to sit high and close to the wheel." Additionally, an optional panic button on the car's key chain would flash the lights and sound the horn if the driver were being stalked on her way to the parking lot. In the case of the Probe, then, a specific vision of womanhood informed not only the marketing, distribution, and consumption of the automobile, but also the very design of the car—what historians of technology often refer to as the "nuts and bolts" of an artifact. Even as gender helped produce a new car design, Ford's Probe subtly crafted a particular embodiment of gender—a "woman" marked first and foremost, of course, by the power to make economic decisions about vehicle purchases.[2] The car also materializes "woman" in other ways: she has the physical ability to operate a car, feet amenable to high heels, hands with long fingernails, legs covered in skirts, a nervous system that tends toward panic, and a mind that seeks strangers' assistance in moments of danger rather than employing other forms of self-defense.

As this example makes clear, specific forms of gender and specific forms of technology are reciprocally constitutive: Probes redesign this "woman" just as a group of particular professional women redesigned the Probe. Other recent studies demonstrate a similar process of mutual constitution working in and around artifacts such as birth control pills and ultrasonography.[3] Through such empirical research, we have learned that adequate histories of technology require complicated, situated histories of gender relations. As the editors of one recent collection on the subject conclude, "It is as impossible to understand gender without technology as to understand technology without gender."[4]

I want to borrow this insight from feminist studies of technology to encourage a similarly subtle and powerful approach to *race* in studies of technology.[5] Cursory surveys of recent scholarship produced by historians of technology reveal that the field lacks subtle attention to issues of race—an oversight that carries consequences for both historiography and politics, as I will discuss in my concluding remarks.[6] Generally, historians of technology ignore the subject of race altogether. Even younger historians of technology, reared on critical historiography and accustomed to social histories of race, have been surprisingly slow to consider the technological constitution of racial identity and difference.[7] When race is broached in mainstream histories of technology, it nearly always appears as an intrinsic, timeless feature of identity, a fixed attribute whose "impact" is then used to explain the use or rejection of specific technologies.

Consider, for instance, an often-reprinted essay titled "From Africa to America: Black Women Inventors," which traces the "creative and inventive abilities" of "black women" from antiquity to the present.[8] Because the essay provides visibility for a number of inventors and inventions often ignored in mainstream histories of technology, it is extremely useful in undergraduate survey courses. At the same time, it neglects critical differences in what "blackness" and "womanliness" have entailed across time and place. "Black womanhood" is instead given a kind of ahistorical essence, unifying creative endeavors from the Nile Valley to the Mississippi Delta, from ancient agriculturalists to modern astronauts.

Overlooked in such approaches, I suggest, are ways in which "race" obtains its very definition through technology. In the antebellum United States, for example, "blackness" was in part circumscribed through legal and extralegal prohibitions on invention, as Portia James elucidates in *The Real McCoy*.[9] Understanding the history of technological innovation in America, then, requires us to move beyond taking race as a timeless feature of human identity, a somatic constant present in inventors since antiquity. Rather, following recent scholarship in critical race theory, we might begin to analyze the formation of particular racial subjects in specific national, cultural, regional, and temporal contexts.

If historians of technology have tended to under-examine race, however, theorists of race have been equally silent on the matter of technology. Even as literary critics, philosophers, and other scholars consider the performativity of race, there has been little explicit attention to the *tools* of race-making, other than the techniques of law (and, by implication, the materiality of weapons, prisons, and transport systems through which law is enforced). Remarkably, this avoidance of technology occurs even in recent considerations of "passing," discussions overtly concerned with the mutability of racial identity.[10] In the narratives of passing analyzed by recent theorists, technologies tend to play a central role in at least two ways. First, they provide the means by which characters establish alternate racial identities (as in John Howard Griffin's 1960 best-seller, *Black Like Me*, in which a white journalist meditates on the artifacts of his alleged race-shifting: "medication" he takes to begin his experiment; "sun lamps" used to augment the tablets' action; razor blades to remove hair from his head and, later, his hands; sponges and Kleenex used to apply and remove skin dye; and, finally, a mirror he uses to shore up his sense of inner "whiteness" in opposition to his new "black" appearance).[11] Second, technologies often offer the key to resolving a character's "true" racial identity in moments of crisis (as when Pudd'nhead Wilson employs the palpable indices of ink, glass, paper, white cardboard, and pantograph enlargements to reveal racially distinct fingerprints in Twain's climactic courtroom scene). The import of these artifacts is rarely mentioned, however, much less analyzed systematically. If one seeks to locate race in time and place, it seems nearly impossible to avoid the various technologies through which U.S. racial categories and identities have been perceived, measured, classified, and altered—to say nothing of how the very concept of "technology" has been racialized in American history.[12]

Why, I invite readers to wonder with me, might we be encountering such a strange gap between these two bodies of scholarship? Why might historians be so able to consider the conjunction of race and sexuality, say, or technology and nation-building, yet remain unable to sustain investigation of the interplay of technology and race? I encounter such questions over and over in the course of teaching, conducting research, and conversing with colleagues. They remain questions for which I do not have adequate answers. In the spirit of this volume's call for provocation, however, this essay will first suggest one way to think about the gap yawning between the two fields of study—a gap that makes sense of two (seemingly) competing intellectual agendas. I will then offer one cursory example of ways to think simultaneously about technological change and the creation of racial subjects. This example is intended to be generative rather that conclusive; I mean only to demonstrate that many other possibilities for further study exist. Finally, the essay will describe some of the dangers of the scholarly lacuna named here, and explain why I think bringing race studies and technology studies closer together matters so much.

Although trends in academic inquiry defy any simple explanation, I speculate that the gap between studies of race and studies of technology emerges in part in the interstices between two separate historical trends: a shift in academic ideas about race since the Second World War, and the disciplinary professionalization of the history of technology. As Nancy Stepan, George Stocking, and numerous others have noted, postwar Western intellectuals beat a hasty, self-conscious retreat from morphological and physical understandings of race—a retreat spurred partly by the triumph of Boasian anthropology, the rise of population genetics, the lingering specter of eugenics, and anti-colonial uprisings. From Ashley Montagu's 1942 analysis of the "fallacy of race" or UNESCO's famous 1948 statement on race and racism, to scientist Frank Livingstone's treatise "On the Non-Existence of Human Races," the assertion that "race does not exist on the biological level" became a kind of mantra for postwar liberal and progressive thinkers.[13] This assertion's functional corollary, that "race is a social construction," endures to this day in nearly all academic fields.[14] The appeal of this idea is not hard to understand. Given that most recent discussions of the physicality of race are linked to overtly or tacitly racist projects (think *The Bell Curve*), one can appreciate a general aversion to speaking about race and anything "physical" in the same breath. We anti-racist scholars have learned to become squeamish about bodies.

In this context, discussions of technologies of race may seem to carry the stench of earlier racist taxonomies—people filling empty human skulls with buckshot to quantify cranial capacities and other, far more ghastly endeavors. Thinking about materiality without implying fixed, natural morphological differences presents difficulty, and since "technology" seems to evoke materiality, the topic is simply avoided altogether. I often encounter this avoidance of race when teaching undergraduate courses at an affluent, private New England college. When I ask students (who self-identify with a range of "races") to imagine ways race and technology might produce one another, most look at me in astonishment. "Race is a social construction, Professor Herzig," they inform me gently, as though someone forgot to take me aside to point that out. For them, "constructions" reside purely in the ethereal realm of beliefs and attitudes; understandably, the "materiality of race" appears as an oxymoron.

Just as postwar intellectual and social movements were working hard to dematerialize concepts of race, a new historical subdiscipline was attempting to rematerialize understandings of technology. From its very incorporation as a professional specialty, the history of technology adopted a meticulous, relentless focus on matter. Indeed, a preoccupation with greasy, sweaty, noisy *physicality* marks one of the primary signs of its disciplinary specificity.[15] As every origin story of the field reports, the former engineers who founded the discipline centered their work on stuff, on the nuts and bolts of our various social worlds. Many historians of technology would still proudly note that what distinguishes us from general historians (or, more importantly, from traditional Sartonesque historians of science) is precisely this attention to the material world—its physical limits, possibilities, and tendencies. Given the peculiar heritage of the history of technology and the turn in thinking about race after World War II, one can perhaps understand a certain squeamishness among historians of technology about addressing race. Can historians of technology address race without naturalizing or essentializing racial difference? By the same token, can theorists of race talk about technology without evading the materiality, the "nuts and bolts-ness," of artifacts and their uses? Is it possible to join the progressive, anti-racist aims of postwar theories of race to the distinctively materialist methods of the history of technology?

I want to suggest that it is possible—that we might build on the institutional, political, and historical commitments of both bodies of inquiry. I will try to demonstrate this claim by mimicking the type of reciprocal analysis established in contemporary feminist studies of technology. My aim here is twofold. First, I want to push our studies of technology beyond viewing "race" as having any stable, self-evident, or natural quality that simply "impacts" technological development and use. I hope that after reading this essay, scholars in science and technology studies might be reluctant to make claims about "black women inventors" from ancient Egypt to contemporary America without articulating strategic reasons for doing so.[16] At the same time, I hope that scholars of race in the United States would begin to take technology as a serious category of analysis, to realize that technologies are never neutral, fixed tools that are simply "adopted" for various racializing projects. Artifacts, like racial subject positions, are subject to constant re-articulation and refashioning. Put most simply, I want to argue that this hazy thing "race" did not exist prior to "technology" nor vice versa; rather, the two emerge simultaneously through particular, identifiable practices. I will illustrate this claim by telling a story, a rather odd story about hair.

In the United States, hair has been a catalyst for discussions of race, technology, nature, and artifice since the founding of the union. Although the great taxonomist Linnaeus presumed that the indigenous peoples of the Americas were naturally beardless, American thinkers—with evident economic self-interest in questions of race—puzzled over the racial and technological attributes of human body hair. Thomas Jefferson, for instance, argued that nature is never quite as natural as it seems, at least where hair is concerned:

Mr. Jefferson remarks that "it has been said that the [North American] Indians have less hair than the whites, except on the head; but this is a fact of which fair proof can hardly be had. With them it is disgraceful to have hairs on the body, they say it likens them to hogs; they therefore pluck it out as fast as it appears. But the traders who marry their women, and prevail on them to discontinue this practice, say, that nature is the same with them as with the whites."[17]

Whereas Jefferson concerned himself with the technological alteration of (perceived) racial characteristics, later commentators such as Peter A. Browne approached technology and hair in a different light. For the antebellum Browne, establishing *distinctions* between races mattered more than ascertaining practices of hair removal.[18] In his elaborate 1853 treatise *Trichologia Mammalium*, Browne delineated "three distinct species of human beings" based on hair types (66). Browne based this controversial polygenist claim on his use of new instruments—the trichometer, the discotome, and the hair revolver—all designed to reveal and quantify somatic hierarchies. The measurement, classification, and manipulation of human body hair, in short, enabled the recognition and classification of "race" in nineteenth-century America.

By the early twentieth century, American race politics had changed once again, and were attended by new or renewed types of technologies of hair and its removal. Many of these technologies—chemical depilatories, safety razors, and electrolysis—remain familiar to readers today. Yet one forgotten technology, x-ray hair removal, provides a particularly clear window into the relations of race and technology in the early twentieth century.

The hair-removing capacities of x rays (given the x in honor of their enigmatic properties) were first noted in March 1896, only four` months after Roentgen's famous announcement of the new kind of light. Two Vanderbilt University researchers discovered the ray's epilating properties when attempting to locate a bullet in the head of a child brought in for treatment. In an effort to determine the feasibility of x-raying the skull, Dr. William Dudley allowed his associate, John Daniel, to experiment on his own head. Twenty-one days after Dudley's skull had been exposed for an hour with the tube placed a half-inch from the scalp, Daniel reported that all hair had fallen out from the area that had been held beneath the x-ray tube.[19]

Within a year of Daniel's report, publications began to appear detailing the application of the rays in the treatment of "hypertrichosis," the medical term for excessive hair. Two Viennese dermatologists, Eduard Schiff and Leopold Freund, are generally credited as the first to undertake methodical experiments in this use of x-rays.[20] Following Schiff and Freund's lead, similarly successful reports emerged from scores of other medical practitioners in both Europe and North America. Medical x-ray epilation enjoyed widespread application through the turn of the century. Even more popular, however, was x-ray hair removal practiced in nonmedical, commercial establishments. The most famous of these ventures was the Tricho Sales Corporation of New York City, founded by physician Albert C. Geyser.[21] By 1925, Geyser's specially constructed "Tricho Machines" were in operation in more than fifty U.S. cities.[22] More than a dozen other similar companies also offered hair removal with the x-ray, marketed under names such as the "magic ray," "light ray," or "Epilax ray."

It is difficult to determine the precise number of individuals who underwent x-ray epilation in these years, although it appears tens of thousands, if not hundreds of thousands, of clients—mostly women—used the process.[23] According to physicians' reports, these clients sought to remove hair from their arms, legs, buttocks, pubic regions, breasts, armpits, or—most frequently—their faces. Medical and legal records indicate that most epilation clients were working women employed in low- or middle-income positions: telephone operators, secretaries, clerks, and so on. Some clients at commercial salons even received special group discounts for bringing in large numbers of friends or coworkers.[24] Records also show that salons directed their x-ray epilation advertisements toward urban, non-English-speaking populations.[25] Medical reports on hypertrichotic patients often remarked on the role of "heredity" in hair growth, particularly in case histories of individuals of Russian or Italian descent. Their discussions of hair and "hereditary derangement" tended to normalize white skin.[26] One practitioner, for example, noted that where epilation had been successful the patient's skin would be "smooth, white, almost normal in character."[27]

Whereas physicians typically took whiteness as a natural fact, the proprietors of x-ray epilation salons understood that hairless, pale skin was an accomplishment, a feat of human *work*. Indeed, the men and women who profited from the practice of x-ray hair removal in the 1920s and the 1930s grasped (and molded) the connections between social and physical labor, body hair, and white racial identity. Amidst anxious interwar discussions of immigration, racial degeneration, and the practice of passing, body hair came to signify both the promise and the threat of slippery racial boundaries. This mood of racial self-transformation is captured in one 1933 brochure, paradoxically titled "Be Your True Self: We will tell you how": "Our MODERN SCIENTIFIC METHOD with filtered RAY" is "not 'just another' hair remover," but the only one that can "banish" one's dark traces. As the brochure pledged, "It is only a step from the Shadow into the Sunshine."[28] The enlightenment offered by scientific knowledge and the woman's own visible, physical "enlightenment" were further entangled in the ads' recurrent visual tropes, including powdered wigs, full skirts, and other markers of eighteenth-century enlightenment culture. The racial overtones of "enlightenment" appear prominently in one salon's letter to a prospective client, which warned: "Whatever you do—wherever you go—you need to have your skin CLEARED from this dark shadow."[29] The technological commodity, created by the "inevitable progress of science," is here suggested as the pathway to personal, physiological transformation.

As scientific advancement was said to yield somatic enlightenment, so too were both narratives of uplift linked to dreams of upward class mobility. Just as salon proprietors implied a connection between race and hair, they also suggested that the establishment of white racial identity translated into enhanced economic opportunity. Through plentiful photographs of plush consultation rooms and sleek laboratories, epilation promoters promised prospective clients a space unequaled in cleanliness and luxury.[30] Simply by stepping into the salon, claimed one testimonial, a woman was invited to experience the "padded carpets, the beautiful polychrome furniture, the soft shaded lights and dainty draperies, [and] . . . the quiet, restful atmosphere of . . . studious habits and refined tastes. . . . [The epilation laboratory is] a vision of sanitary loveliness. The place is done in pure white enamel, not an atom of dust can enter here."[31]

Many of these promotions made the economic importance of technological self-fashioning far more explicit. Prefaced by a line drawing of a bourgeois white couple approaching a cap-wearing attendant, a Chicago Marveau Laboratories ad demanded, "Can you afford to neglect your personal appearance any longer, in this age when it counts so much in social and economic advancement?"[32] Similarly underscoring the sexual (and hence, economic) benefits of feminine hairlessness, the Dunsworth Laboratories of Indianapolis seconded this theme with their statement: "Freedom from Unwanted Hair Opens the Gates to Social Enjoyments that are Forever Closed to Those so Afflicted."[33]

Removing the "dark shadow" that barred access to the world of "social enjoyments" appears to have acquired a certain urgency during the interwar period, an era of increasingly restrictive U.S. immigration laws, widespread racist violence, growing interest in eugenics, and deepening economic depression.[34] We find expressions of this anxiety in letters composed by epilation clients. One woman summarized the mood of these letters when writing to the AMA in 1931. Under the headline of a brochure from Philadelphia's Cosmique Laboratories, Katherine Moore asked simply, "Doesn't this sound pretty good?"[35] A 25-year-old Brooklynite wrote to the American Medical Association in 1933 to name her concern:

I am working for quite a small salary now and I have saved every penny I possibly could. Denying myself luxuries that all girls love toward trying to get rid of this unwanted hair, and believe me that saving this money was quite an effort as things at home are quite bad. But, I cannot go on as I am now, as I am miserable through a freak of nature and I have more than once thought of putting an end to my misery.[36]

Like other members of the nation's urban working poor, this young woman could scarcely afford to ignore the possibility of economic uplift seemingly carried by what the ads termed a "complexion clear and fair."[37]

What was at stake in x-ray epilation, I would suggest, was the technological fashioning and refashioning of white racial identity, a process occurring in a moment of intense reordering of received racial categories.[38] Hairs that marred and troubled the boundaries of the "white female" body were expulsed, rendered abject by the invisible, sanitizing light of the x-ray.[39] In other words, this tool formed the very substance of the raced, sexed, subject. Technology re-shaped "race" in this instance not only symbolically (through epilation advertisements), but also *materially*—indeed, at the level of the irradiated, cancerous cell.

Even as technologies were being used to reshape race, race was reshaping technologies, both symbolically and materially. "The" x-ray was no neutral, stable, technical artifact, one unfortunately appropriated by a few quacks; x-ray epilation does not offer a tale about the misapplication of an essentially apolitical object. Rather, the technology itself was reconstituted through these practices of race-making. The redesign of the x-ray at once facilitated and was facilitated by the shifting significance of "race" in the interwar period, a process made clear by closer examination of the work of Albert C. Geyser.

Moved by colleagues' unsuccessful treatments of hypertrichosis and patients' pleas for help, Geyser turned his attention to the problem of "excessive hair" in the opening years of the twentieth century.[40] By 1905, Geyser developed the Cornell tube, named for the college with which he was then associated.[41] The innovation behind the Cornell tube, a more or less standard version of the then-common cathode ray tube, was the use of upgraded materials. Made of heavy lead glass through which no rays were thought to be emitted, the tube contained at one end a small, permeable flint glass window that was brought into direct contact with the skin. Geyser claimed that the lead filtration eliminated all unsafe radiation, and he later insisted that in over 5,000 applications of the tube, "not a single case of [radio] dermatitis had developed."[42] The Cornell tube made Geyser quite famous, and his 1908 wedding made the front page of the *New York Times*.[43]

Even as Geyser's Cornell tube was revolutionizing the practice of women's hair removal, the technology also left its mark on Geyser's own body. Like most early x-ray experimenters working without dosimetry equipment, Geyser often used his own hands as "measuring instruments" when testing his machines. As a result of such prolonged, repeated exposures to radiation, Geyser eventually surrendered all of the fingers, metacarpal bones, and one row of carpal bones on his left hand to arrest the spread of x-ray-induced cancer. Another 24 ulcers were excised from his right hand, leaving the physician permanently maimed.[44] Driven by the loss of his hands, Geyser continued his research into new and improved technologies of hair removal, focusing, not surprisingly, on ways to automate the process.[45] In conjunction with the Wappler Electric Manufacturing Company of New York City, Geyser perfected a special apparatus (his "Tricho Machine") to administer a unique wavelength of ray, defined as 1/30,000,000 of a millimeter.[46] Geyser's greatest interest was in standardizing and automating the dosage of radiation provided by the apparatus, thus preventing irregular, dangerous exposures and largely eliminating the need for manual operation. As Geyser described the newly mechanized epilation procedure:

When a patient presents herself or himself for treatment, a metal applicator, the shape of the area to be treated, is adjusted to the machine. The patient is seated and the particular hair surface is exposed through the applicator to the Tricho wave. . . . The mere turning of a switch starts the operation and the machine automatically shuts off after the proper time.[47]

The machine required three to four minutes of exposure for effect, repeated once every two weeks. The patient's hair was supposed to fall out after five to seven treatments.[48]

Unlike the Cornell tube's uneven and prolonged exposures, this new machine delivered Geyser's "dream of faultless skin" with robotic precision. Geyser repeatedly emphasized the homogeneity and fixity offered by the Tricho apparatus. The voltage, distance from target, and length of exposure were all fixed in accordance with Geyser's aim of making "faultless skin."[49] Just as the very nature of "whiteness" emerged in and through different instantiations of x-ray epilation, so too a different kind of x-ray emerged in and through these racialized and racializing practices. New tubes, new materials, new casings for the equipment, new switches and levers, new protective equipment, new metal applicators, and new patents all emerged here in relation to the shifting role of "race" in the interwar years.

Attention to the history of race in the United States thus alters our understanding of the history of technology. Rather than attributing developments in early-twentieth-century x-ray technology simply to radiologists' increasing design skills or to their increasing mastery of materials such as glass and tungsten, I am suggesting that early American radiology must also be considered in relation to the permutations of the period's fraught and complicated race relations. New tools emerged in the context of specific politics of race. Given that hair was not the only sign of race in the interwar period, it should come as no surprise to learn that several experimenters adapted x-ray machines to attempt x-ray skin lightening.[50] In short, just as race relations necessitate attention to the tools of race-making and race-changing, so, too, our histories of

American technology require analysis of the ways race has informed the development, distribution, and use of particular artifacts.

I have suggested that there is much to be gained by combining the analytical strengths of two fields of recent scholarship. Historians of technology, despite their keen interest in and attention to the materiality of historical subjects, have tended to avoid questions of race. Historians and philosophers of race, despite their attention to the production and maintenance of racial identities, have had remarkably little to say about the significance of technology in this process. Just as "race" has played little role where historians of technology are concerned, so has "technology" played little role where theorists of race are concerned. My aim in this essay has been to draw attention to the fact that technologies have been used to materialize racial identities and racial differences, even as race gives concrete, material form to particular technologies. This is, of course, hardly a novel theoretical intervention: not only does a whole tradition of feminist research demand that we carefully consider the *matter* of bodies, but whole fields of research in science and technology studies urge consideration of the "coproduction" or the "relational materiality" of artifacts and users.[51] I hope merely to spark additional conversation between two still-distant fields. My intention is to show the methodological importance of moving beyond the "impact" school of analysis—to approach race and technology as things-in-relation.

But why does this methodological move matter? Why should we bother to bridge these two fields? In closing, let me try to explain what I see as at stake here. I would suggest there are at least two reasons for seeking to join these two fields.[52] First, changing our ways of thinking about race and technology allows us to ask new, and critical, questions. Put simply, it provides better history, for new approaches allow us to ask new questions about the past. What role did tools such as the "spirometer" (designed to ascertain "Negro" lung capacity) play in fixing or transforming antebellum attitudes toward slavery and abolition? How did the development and use of instruments such as callipers, cephalometers, craniometers, and parietal goniometers inform Indian removal policy in the United States, or the rise of Chinese exclusion laws, or the persistence of anti-miscegenation laws?[53] How have specific tools—rhinoplastics, hair straighteners or curlers, skin lighteners or darkeners—been used to subvert or overthrow existing racial exclusions and violence? How has "technology" figured in the revision of "scientific" ideas of race, including that much-discussed "retreat from race" in the 1940s? As attention to women's lives opened entire fields of historical knowledge, so, too, bridging race and technology brings new questions and answers into view.

A second reason for rethinking the relations between race and technology lies not in the realm of historiography but more firmly in the realm of politics. Arguments that race is a "social construction" (and hence immaterial) allow us to ignore the material persistence of race in contemporary society, particularly in our sciences and technologies. In contemporary forensic anthropology, in *Sports Illustrated*'s recent cover story about the lamentable demise of the white athlete, in the much-hyped Human Genome Project, and in numerous other locations, we see the re-emergence of insidious conceptions of race, race suicide, and eugenic social policy, now veiled from our critical scrutiny by the deafening refrain that "race doesn't exist on the biological level."[54] How are we to address such phenomena without subtle, sophisticated analyses of the culturally and historically specific matter of raced bodies? How are we to address disproportionate infant mortality rates, differing susceptibilities to hypertension or anorexia or fetal alcohol syndrome without more subtle, sophisticated analyses of the materiality of "race"? How are we to resist the ways racial classification functions to restrict economic and social resources, re-entrenching wealth and status among some? In the words of Donna Haraway: "Race kills, liberally and unequally; and race privileges, unspeakably and abundantly."[55] Race is not unvarying, to be sure. It is not fixed on the skin, or in the genes. But at the same time, like all aspects of human identity, race *is* embodied, and the ways in which it is embodied can cause more or less physical pain, more or less death. If we are to resist the unequal distribution of suffering, then, it seems imperative to keep the matter of race firmly in our historiographical sights, and keeping an eye on technology seems a particularly productive way to do so.

Acknowledgments

This work was originally prepared for a colloquium presentation at Colby College to the Programs in Women's Studies and in Science, Technology, and Society. I am grateful to Elizabeth D. Leonard, Jenny Reardon, Lenny Reich, Carroll Pursell, and Bruce Sinclair for critical comments and warm encouragement.

Notes

1. See David J. Hess, *Science and Technology in a Multicultural World: The Cultural Politics of Facts and Artifacts* (Columbia University Press, 1995), p. 44–45; Mimi Vandermolen, "Shifting the Corporate Culture," *Working Woman*, November 1992: 25–28.

2. The term "gender" here raises the specter of a massive debate ongoing within feminist theory and women's studies over the distinctions (such as they may be) between sex and gender—a debate that engages all the questions of materiality and power I seek to address in this essay. A helpful synopsis of some of these debates may be found in the special issue of *Australian Feminist Studies* 10 (1989), although these essays far from exhaust the subtleties in the discus-

sion in recent years. I refrain from defining "gender" (or, for that matter, "race" and "technology") here precisely because I think that the debates themselves are more useful than simple definitions. I would concur, however, with the several historians of technology who have named gender as, above all, a tool for further thought: gender, they write, "is not just a fancy word for women's history. Studying gender and technology is not a question of counting female engineers, although that information may well be pertinent. Rather, gender is an analytical tool useful for making sense of culture, and thus for exploring the relationship between culture and technology." See Nina E. Lerman, Arwen Palmer Mohun, and Ruth Oldenziel, "Versatile Tools: Gender Analysis and the History of Technology," *Technology and Culture* 38 (1997), no. 1: 2–3.

3. Nelly Oudshoorn, *Beyond the Natural Body: An Archaeology of Sex Hormones* (Routledge, 1994); Karen Barad, "Getting Real: Technoscientific Practices and the Materialization of Reality," *differences* 10 (1998): 87–128. See also Carroll W. Pursell, "The Construction of Masculinity and Technology," *Polhem* 11 (1993): 206–219.

4. Nina E. Lerman, Arwen Palmer Mohun, and Ruth Oldenziel, "The Shoulders We Stand On and the View from Here: Historiography and Directions for Research," *Technology and Culture* 38 (1997), January, p. 30.

5. I begin this essay with examples from feminist studies of technology simply because they make technology studies' concerted silence about race more plain. My thinking on race and technology has been equally shaped by feminist theorists of race, particularly Toni Morrison (*Playing in the Dark*, Harvard University Press, 1992) and Evelyn Brooks Higginbotham ("African American Women's History and the Metalanguage of Race," *Signs* 17, 1992, winter: 251–275).

6. Although it is impossible to limit "the field" in any clear way, I am here characterizing history of technology as represented by work presented to recent meetings of the Society for the History of Technology (founded in 1958) and the International Committee for the History of Technology (founded in 1968), and in various leading journals such as *Technology and Culture*, *History and Technology*, and *ICON*. My discussion also focuses on scholarship addressing technology in the United States, as befits the titular subject of this volume.

7. Even the most critical re-assessments of race and technology typically ignore material re-formations of racial identity and difference. For two crucial exceptions to this trend, see Carlos E. Martín, "Mechanization and 'Mexicanization': Racialization and California's Agricultural Technology" (unpublished manuscript) and Evelynn M. Hammonds, "New Technologies of Race," in *Processed Lives*, ed. J. Terry and M. Calvert (Routledge, 1997).

8. Autumn Stanley, "From Africa to America: Black Women Inventors," in *The Technological Woman*, ed. J. Zimmerman (Praeger, 1983), p. 63.

9. Portia P. James, *The Real McCoy: African-American Invention and Innovation, 1619–1930* (Smithsonian Institution Press, 1989). See also her essay in this volume.

10. Noteworthy exceptions (though still, perhaps, unsatisfactorily "technical" in the eyes of many mainstream historians of technology) include Gayle Wald, "'A Most Disagreeable Mirror: Reflections on White Identity in *Black Like Me*," in *Passing the Fictions of Identity*, ed. E. Ginsberg (Duke University Press, 1996); Susan Gillman, "'Sure Identifiers': Race, Science, and the Law in *Puddn'head Wilson*," in *Mark Twain's* Puddn'head Wilson, ed. S. Gillman and F. Robinson (Duke University Press, 1990); Michael Rogin, "Francis Galton and Mark Twain: The Natal Autograph in *Puddin'head Wilson*," in the same volume.

11. John Howard Griffin, *Black Like Me* (Signet, 1976 [1960]), pp. 12, 120, 28, 117, 124, 15.

12. In very different ways, Michael Adas and Leo Marx provide starting points for a race-conscious history of the term "technology." See Adas, *Machines as the Measure of Men* (Cornell University Press, 1989) and Marx, "The Idea of 'Technology' and Postmodern Pessimism," in *Does Technology Drive History?* ed. M. Smith and L. Marx (MIT Press, 1994). An additional starting point for the consideration of the racialization of technology can be found in Gyan Prakesh's discussion of Kipling in *Another Reason: Science and the Imagination of Modern India* (Princeton University Press, 1999), p. 5. Recent attention to the racial, economic, and sexual density of discourses of "civilization" should alert us to the similar implications of "technology"—one of the key analytical categories employed by nineteenth-century taxonomists in their evaluations of relative advancement and degeneracy.

13. In a provocative essay, Kate Baldwin argues that this liberal affirmation of the biological similitude among races veiled a larger anticommunist program. See "Black Like Who? Cross-Testing the 'Real' Lines of John Howard Griffin's *Black Like Me*," *Cultural Critique* 40 (1998), fall: 103–143.

14. Despite the best efforts of theorists such as Judith Butler and Karen Barad to recast "construction" in terms more subtly aware of the agency of the material world, theories of race have tended to adopt scientists' strict biology/ideology bifurcation. Indeed, leading cultural theorists often explicitly credit "science" for revealing the "constructedness" (i.e., material fictitiousness) of race. See, for instance, Paul Gilroy, *Against Race: Imagining Political Culture beyond the Color Line* (Harvard University Press, 2000).

15. See, e.g., John M. Staudenmaier, "What SHOT Hath Wrought and What SHOT Hath Not," *Technology and Culture* 25 (1984): 707–730.

16. Such strategic reasons are not difficult to imagine, in the context of twenty-first-century racial and economic inequities. For example, Jenny Reardon discusses some of the self-consciously essentialist racial narratives adopted and advanced by opponents of human genetic sampling initiatives (dissertation, Cornell University, 2000).

17. Peter A. Browne, *Trichologia Mammalium; or, A Treatise on the Organization, Properties and Uses of Hair and Wool* (Philadelphia, 1853), p. 31.

18. See Peter A. Browne, *The Classification of Mankind, by the Hair and Wool of their Heads, with the Nomenclature of Human Hybrids* (Philadelphia, 1852); William Stanton, "A Perfect Hair," in *The Leopard's Spots: Scientific Attitudes toward Race in America, 1815–59* (University of Chicago Press, 1960).

19. John Daniel, *Science* 3 (1896): 562–563, as quoted in Ruth Brecher and Edward Brecher, *The Rays: A History of Radiology in the United States and Canada* (Williams & Wilkins, 1969), pp. 81–82.

20. Eduard Schiff and Leopold Freund, "Beiträge zur Radiotherapie," *Wiener Medicinische Wochenschrift* 48, no. 22 (1898): 1058–1061.

21. Geyser's own account of his professional development can be found in a letter written to Dr. Fischbein, editor of the *Journal of the American Medical Association*, October 29, 1925. American Medical Association Historical Health Fraud Collection, Chicago, Illinois (hereafter AMA), folder 0318-03.

22. Reprint from *The Business Survey*, November 25, 1925, AMA folder 0318-01.

23. In terms of total numbers of clients overall, two physicians writing in 1947 estimated that the "cases of x-ray burns, cancer and death resulting from treatments administered by the Tricho

Institute must have run into the thousands," from which we might speculate that the total number of women and men receiving the treatment was even higher. A. C. Cipollaro and M. B. Einhorn, "Use of X-Rays for Treatment of Hypertrichosis Is Dangerous," *Journal of the American Medical Association* 135 (1947), October 11, p. 350.

24. Edward Oliver, "Dermatitis Due to Tricho Method," *Archives of Dermatology and Syphilology* 25 (1932), p. 948; D. E. Cleveland, "The Removal of Superfluous Hair by X-Rays," *Canadian Medical Association Journal* 59 (1948), p. 375.

25. Herman Goodman, M.D., "Correspondence," *JAMA*, May 9, 1925, p. 1443; S. Dana Hubbard, M.D. (City of New York Dept. of Health) to A. J. Cramp, January 8, 1929, AMA folder 0318-02.

26. Dr. Gilmour, "Hypertrichosis," *Journal of Cutaneous Diseases* 36, no. 4 (1918), p. 255.

27. Dr. Heidingsfeld, "Hypertrichosis," *Journal of Cutaneous Diseases* 36, no. 5 (1918), p. 308. Compare Albert C. Geyser, "A Résumé of Two Hundred Cases of Hypertrichosis Treated with the Roentgen Ray," *Journal of Cutaneous Diseases Including Syphilis*, July 1915, p. 3.

28. "Be Your True Self: We Will Tell You How" [Advertisement for Frances A. Post, Inc., Cleveland], AMA folder 0317-01.

29. H. Gellert [Secretary of Hamomar Institute] to "Madam," 1933, AMA folder 0317-01.

30. Copy of letter, Miss H. Stearn to National Institute of Health, July 5, 1933, AMA folder 0317-01; "BEAUTY: Woman's Most Precious Gift" [Advertisement for the Dermic Laboratories, San Francisco and Los Angeles], AMA folder 0317-02.

31. M. J. Rush, "Hypertrichosis: The Marton Method, A Triumph of Chemistry," *Medical Practice*, March 1924, p. 956, reprint in AMA folder 0317-01.

32. "Beauty Is Your Heritage" [advertisement for Marveau Laboratories, Chicago], AMA folder 0317-02.

33. "Permanent Freedom from Unwanted Hair," AMA folder 0317-02.

34. My perspective on early-twentieth-century eugenics has been most influenced by Donna Haraway's "Teddy Bear Patriarchy: Taxidermy in the Garden of Eden, New York City, 1908–1936," in Haraway, *Primate Visions: Gender, Race, and Nature in the World of Modern Science* (Routledge, 1989).

35. Katherine Moore to AMA, August 27, 1931, AMA folder 0318-02.

36. Anne Steiman to AMA, September 5, 1933, AMA folder 0317-01.

37. "A Flawless Skin," *Boston Post*, November 15, 1928, copy in AMA folder 0318-02.

38. Peggy Pascoe, "Miscegenation Law, Court Cases, and Ideologies of 'Race' in Twentieth-Century America," *Journal of American History* 83 (1996), June: 44–69; Ian F. Haney López, "White by Law," in *Critical Race Theory*, ed. R. Delgado (Temple University Press, 1995).

39. Judith Butler discusses a similarly tenuous, repetitive materialization of whiteness in her book *Bodies That Matter: On the Discursive Limits of "Sex"* (Routledge, 1993), p. 275, n. 4.

40. Albert C. Geyser, "The Choice of Methods between the Use of the Roentgen Ray and the High-Frequency Currents in Therapeutics," *Journal of Advanced Therapeutics*, June 1907, reprint at New York Academy of Medicine (38532).

41. Albert C. Geyser, "Using the X-Ray without Burning," *JAMA* 1, no. 13 (1908): 1017–1019, reprint at New York Academy of Medicine (41340), p. 3.

42. Ibid., p. 3.

43. "X-Ray Expert Wins a Bride," *New York Times*, April 16, 1908.

44. Geyser to Fischbein, October 29, 1925, AMA folder 0318-03. It is worth noting here that the grueling experience of the (white, male) physician materializes race in a particular way, as well. For more on the relations between suffering and the formation of the medico-scientific subject, see Susan W. Hinze, "Gender and the Body of Medicine, Or At Least Some Body Parts: (Re)Constructing the Prestige Hierarchy of Medical Specialties," *Sociological Quarterly* 40 (1999), no. 2: 217–239; Rebecca M. Herzig, In the Name of Truth: Sacrificial Ideals and American Science, 1870–1930, Ph.D. dissertation, Massachusetts Institute of Technology, 1998.

45. "Facts and Fallacies about the Removal of Superfluous Hair," AMA folder 0317-17.

46. Geyser's working relationship with Wappler Electric grew along with his interest in hair removal (see Geyser, "The High-Frequency Current in Therapeutics," *New York Medical Times*, July 1910, back cover; Geyser, "The Physics of the High-Frequency Current," *New York Medical Journal*, November 4, 1916, p. 7. Wappler Electric had its hand in several interesting lines of production at this time; for further discussion, see Rachel Maines, "Socially Camouflaged Technologies: The Case of the Electromechanical Vibrator," *IEEE Technology and Society Magazine*, June 1989: 3–23, especially n. 4 on p. 9.

47. "Facts and Fallacies about the Removal of Superfluous Hair," AMA folder 0317-17.

48. Ibid.

49. Geyser, "Hypertrichosis and Its Treatment," *Urologic and Cutaneous Review* 27 (1923); reprint in AMA folder 0318-01.

50. See, e.g., "All Coons to Look White: College Professors Have Scheme to Solve Race Problem," *New York City Morning Telegraph*, January 22, 1904, clipping in box 59, folder 2; "Radium and X-Ray Used to Beautify," *Boston Herald*, May 8, 1904, clipping in box 59, folder 3; "'Can the Ethiopian Change His Skin or the Leopard His Spots': Radium Light Turns Negro's Skin White," *Boston Globe*, January 25, 1904, box 60, folder 2; "Burning Out Birthmarks, Blemishes of the Skin and Even Turning a Negro White with the Magic Rays of Radium, the New Mystery of Science!" *New York American*, January 10, 1904, box 60, folder 2; all in William J. Hammer Collection, 1874–1935, Archives Center, National Museum of American History, Washington DC.

51. For a particularly useful consideration of materiality in science and technology studies (though it ignores the decades of work already completed by feminist theorists of performativity and their critics), see John Law, "After ANT: Complexity, Naming and Topology," in *Actor-Network Theory and After*, ed. J. Law and J. Hassard (Blackwell, 1999).

52. These two reasons echo philosopher Jean-François Lyotard's description of the "speculative" and "emancipatory" narratives for the legitimation of knowledge. See his book *The Postmodern Condition: A Report on Knowledge* (University of Minnesota Press, 1984 [1979]).

53. James D. Guillory, "The Pro-Slavery Arguments of Dr. Samuel Cartwright," *Louisiana History* 9 (1968), p. 221; Nancy Stepan, "Race and Gender: The Role of Analogy in Science," in *Feminism and Science*, ed. E. Keller and H. Longino (Oxford University Press, 1996), p. 125.

54. See, e.g., S. L. Price, "Is It in the Genes?" *Sports Illustrated* 87 (1997), December 8: 52–55.

55. Donna Haraway, *Modest_Witness@Second_Millennium* (Routledge, 1997), p. 213.

Minority Engineering Education in the United States since 1945: A Research Proposal

Amy Slaton

Any panoramic vista of the history of the United States since 1945 will reveal two broad cultural patterns: a tremendous expansion of the nation's technological capacities and the growth of civil rights consciousness across ethnicities and genders. Seen independently, these transformations appear as almost epochal achievements, each fulfilling some of the most progressive promises of an "American century." But no sooner do we identify these two currents of change than their intersection, or rather, their lack of intersection, presents itself as a puzzle. The social profile of American science and engineering occupations, today as in 1945, reflects not a proportionate cross-section of the country's population but rather a severe under-representation of black, Hispanic, and most other minorities. This exclusion of most non-white practitioners from the upper echelons of technical disciplines has been addressed from statistical and policy standpoints for almost five decades and some correctives have emerged. Yet the disparity in educational and job opportunities for non-white scientists and engineers persists. How is it, exactly, that large parts of a culture otherwise committed to the new—in fact, those very cultural enterprises seemingly most available to intellectual innovation— remain in this regard socially backward? This essay explores the social intransigence that has characterized America's technical modernization.

I am proposing here a study that will provide a historical backdrop for some of the quantitative "snapshots" that we have accumulated of individual institutions and policy responses to racial inequities in the sciences, focusing at this time on minority engineering education in the United States since 1945.[1] By any measure, African-Americans and members of other non-Asian minorities remain under-represented among engineering graduates today despite decades of governmental and institutional efforts at diversification. There are few narratives that trace long-term racial patterns in American technical education or the bases of these patterns in the country's changing racial and legal environment.[2] Such a contextualization would help us understand both successful and unsuccessful attempts to integrate engineering and the absence of such attempts altogether in certain settings. Using archival materials, artifacts, and oral history interviews, the proposed study addresses minority engineering education during four historical episodes: the science and engineering "manpower" crisis during and immediately after World War II; the rising tide of integration in the 1950s as separate-but-equal doctrines fell away; the coalescence of civil rights activism and legislation in the 1960s and the 1970s; and the ascendance of political conservatism and a "post-industrial" economy in the 1980s and the 1990s. Within each era I will pair engineering programs of largely black schools with those of predominantly white schools in the same state to compare technical training in the two types of venues, choosing pairs that illustrate the salient cultural and political features of each period.

Two important caveats may be worth stating plainly. First, the experiences of African-Americans in higher education are of course not identical to those of Hispanic, Asian and other minority populations, but they do offer a suggestive starting point from which comparisons and differentiations may be drawn. Second, in comparing traditionally black and traditionally white schools, we should not expect to find a universal deficiency in engineering programs of the former nor a monolithic segregationist impulse in the latter. To presume the existence of either phenomenon would elide the varied attitudes of individual educators and the broad historical forces with which universities have contended since 1945. The very fact of a changing racial and legal environment in America over the second half of the twentieth century makes the glacial pace of integration in technical education particularly vexing for historians because it immediately subverts any easy cultural explanations. Again: Why has engineering as a whole remained largely immune to the reformist forces to which other academic disciplines, and of course many individual engineers and educators, seem subject? One helpful strategy in answering such questions may be to study not simply educational institutions and policies as they have emerged over the last sixty years but also the *content* of technical education: the bench-level practices deployed in engineering laboratories and classrooms. Scholars in science and technology studies (STS) now recognize that the activities of scientists and engineers represent not simply practical bodies of knowledge but also systems of social values shared among practitioners and their audiences.[3] By probing the seemingly mundane daily operations of instruction for black and for white engineering students, we can locate mechanisms of occupational exclusion.

For example, analysts have assessed and improved the efficacy of mentoring, remediation, and other such interventions in minority science education in recent decades. However, the degree to which engineering programs actually prepare students for employment, and for what level of academic or industrial position, can be clarified by examination of the curricula, equipment, and exercises that constitute a given program. Clearly, in any era modernity is crucial: If students receive more advanced training, in more up-to-date facilities, they will emerge better qualified. In this project I will ask how those who have taught different populations of students, and those who have delegated resources for minority and non-minority engineering education, have addressed the material demands of a discipline with ever-changing definitions of cutting-edge practice.[4] Importantly, engineers' career achievements have not necessarily correlated only with more obviously high-tech skills—subjective faculties of intuition and observation, and a cultivated appreciation of arts and letters, have long been associated with technical programs intended to produce research or managerial personnel. In my previous work on materials engineering in the United States between 1900 and 1930, I traced the association of high-level skills and career achievement with white, male practitioners to the presence of such "extra-technical" features of university engineering programs.[5] I will therefore ask of the programs under study here: In what proportion are liberal arts and technical courses included? Are tasks in technical courses idealized or abstracted as would be scientific work, or more strictly imitative of industrial conditions? Does one or the other of these two approaches seem to better prepare engineering students for successful careers? In the same vein, we might ask: In what proportion are the materials presented theoretical or practical? Even such details as the nature of student reports and the calibration of laboratory instruments can reveal educators' conceptions of necessary skills or of "innate" capacities of different student populations.[6]

In the interpretive techniques of science and technology studies, bodies of knowledge formulated and transmitted among social cohorts are studied through texts, instruments, and other materials that codify and disseminate scientific or technological practices.[7] In engineering classrooms since 1945, instructors and administrators have supplied technical knowledge in certain forms and quantities; these distributions of knowledge have shaped the occupational opportunities of students. At the same time, occupational hierarchies and societal organizations in general—including those of racial difference—have shaped activities within university engineering programs. Simply put, an STS approach allows us to identify who, in each engineering classroom or laboratory, is intended to learn what, and why. In creating a detailed portrait of black and white experiences in post-secondary engineering programs the study will contribute to our knowledge of how science has been taught over the last fifty years; how American universities have addressed issues of racial imbalance; and how engineering and modern structures of social opportunity are intricately and perhaps inevitably intertwined. [8]

Case Studies

1945–1954

Approaching post-secondary technical education since 1945 as both a reflection of prevailing social ideologies about race and a shaping force of those values, I have chosen for this study four pairs of schools—one pair for each era—that represent the shifting circumstances of American engineering education after 1945, and that in a general way reflect regional differences among institutions of higher learning. The study will focus first on engineering programs at the University of Missouri, in Columbia, Missouri, and at the historically black Lincoln University, in Jefferson City, Missouri. As land-grant schools founded in the later nineteenth century, both universities represented commitments to practical, rather than classical, education, and both entered the later 1940s with an acute sensitivity to conditions under which the nation educated its technical personnel following World War II. Race-based differences in post-secondary engineering programs become immediately apparent in this comparison, with Lincoln facing not only long-standing political obstacles to its growth and development but also disappointing failures of emergent liberal impulses.[9]

Before World War II, the few technical programs open to black students had been overwhelmingly focused on preparation for the trades, rather than for engineering professions.[10] This heritage of "industrial education" persisted in black land-grant schools, chartered by the Second Morrill Act of 1890, and in private schools such as the venerable Hampton Institute.[11] Howard University alone maintained a full-scale engineering program, established in 1914.[12] Howard graduated increasing numbers of black engineers for many decades thereafter, but for the purposes of this study it is the school's exceptionalism even at mid-century that is telling. Although World War II led many Americans to seek an expanded pool of technical experts and to acknowledge the essential incompatibility between segregationist and anti-Fascist ideologies, and both of these trends gave rise to a rhetoric of commitment to upgrading minority engineering opportunities, little action of that kind was taken.

Lincoln University, sensitive to new manpower concerns, sought to upgrade its engineering offerings to levels established at its almost entirely white counterpart, but rapidly encountered difficulties that reveal the imbalance of resources with which black schools entered the second half of the century, and the degree to which white-dominated political systems controlled the operations of black public institutions.[13] During the war, non-minority public universities received substantial outlays in federal funds for technical programs, and the University of Missouri organized extensive war-related engineering training projects that shaped the school's post-war programs.[14] American industry had little interest in hiring blacks for skilled positions, even at the height of the war emergency, offering a justification to some politicians for denying public funds to black schools. In addition, organized labor in this period firmly objected to the use of public monies for minority technical education.[15] Neither patriotic fervor nor economic desperation significantly displaced racial bias in this period, as the experiences of Lincoln University will show.[16]

The form and efficacy of integrationist arguments in the sciences are central to this inquiry and two rhetorical patterns are brought to light by the cases of Lincoln and the University of Missouri. First, arguments that the inclusion of black Americans in technical professions will benefit the general economic or military status of the nation are still common today, but we might ask why, in any era, arguments for equal opportunity are not based on general notions of civil rights or democratic process.[17] Perhaps fiscal or security arguments for racial integration have constituted a deliberate appeal by their advocates to conservative sectors of society, invoking an indisputable collective good that might transcend racial bias. Whether supporters of black engineering education, such as those at Lincoln, were successful with their arguments or not can tell us a great deal about their audiences, and how prevailing value systems surrounding science or social equity yield to change.

Engineering has long had a powerful symbolic role in the promotion of American social movements and in the 1940s, as in the early 1900s, such rhetorical formulations could represent fairly subtle strategies for social change. If liberal rhetoric at the turn of the twentieth century sometimes masked conservative social agendas, the opposite may have been the case here. Some advocates of improved technical education for blacks at Lincoln and elsewhere celebrated no less lofty a concept than the propensity of science and engineering automatically to carry "civilization forward," and thus the advisability of drawing in more practitioners to these fields. This seemingly ironic denial of the systematic social bias displayed by these enterprises may actually have offered Lincoln's supporters a strategic means of enlisting support for their proposals.[18] The reputation of science as a "value neutral" or even democratic pursuit, while patently unfounded in these cases, may nonetheless have been worth perpetuating. Through study of episodes such as this one we may help historicize the conventional American characterization of science and engineering as impartial, apolitical enterprises—itself a characterization with political implications.[19]

1954–Early 1960s

The early 1950s brought a general atmosphere of increasing civil rights awareness as well as formal objections to segregation in the form of lawsuits against state universities.

As the launching of Sputnik spurred increased fear of technical manpower shortages in the U.S., a few "pre-equity" programs brought technical training to Americans previously denied such opportunities.[20] However, these developments did not dismantle endemic racism, but instead prompted resolutely white state schools to commit further resources to segregation. Some state university systems upgraded their black branches in compliance with separate-but-equal laws rather than face state-mandated admission of black students, yet neither these gestures nor scattered cases of integrationist progress brought an upturn in the number of black engineering graduates.[21] The *Brown v. Board of Education* decision of 1954 that rendered separate-but-equal practices unlawful thus entered an uncertain climate; as schools located in a "border" state, the University of Maryland and Maryland State College (formerly known as the Princess Anne Academy; now known as University of Maryland-Eastern Shore) illustrate these tensions well.[22]

The multi-campus University of Maryland (UMD) displayed in the 1950s a clear commitment to technical advancement, lavishing funds on its physical plants and engineering programs. Ambitious programs of continuing education at the school's College Park and Baltimore campuses and related extension services prepared many returning GIs for technical employment.[23] As well, university directors intended these extension programs to "allay Negro demands to enter the university proper."[24] It is unclear as written whether this action represented an authentic effort at diversification or a minimal attempt to quash black activism; one suspects the latter. UMD had long shown an interest in keeping black students at a distance from the main campuses (where, President Henry Byrd had said in 1937, "our girls are").[25] Maryland State College (MSC) had been formally incorporated into the University of Maryland system in 1935, and some state leaders had moved to supply funds to all "Negro branches" of the university for exclusionary purposes. However, appropriations for those branches were always very small, and they stayed so even after *Brown*.[26] While expanding its College Park and Baltimore technical programs in the late 1950s and early 1960s, the University of Maryland saw little reason to do so at the Eastern Shore location, apparently capitulating to MSC's heritage as a public "normal school" intended to educate black teachers in the tradition of many historically black colleges and universities (HBCUs). Since the post-Civil War period, such programs had elevated the general black population by providing trained primary and secondary school teachers, but they also limited the professional options of black college attendees. [27] In 1962, the University of Maryland's president explained of MSC that the "educational standards [there] cannot be raised above the level of preparation of the majority of students."[28] It is possible that this statement did not reflect a racial criteria for his administrative decisions, but whatever the president's own motivations may have been, MSC faculty were well aware that—like arguments about the limited employment opportunities for black graduates—such statements justified and perpetuated racial inequities.[29] I will examine Maryland's amalgam of technical ambition and social conservatism as it shaped engineering instruction at both state schools.

Later 1960s–1970s

Throughout the 1960s and the 1970s, Affirmative Action plans, outreach programs, and improved enforcement of federal civil rights laws encouraged the admittance of black students to formerly segregated schools and the establishment of many "targeted" programs for minority students in science and engineering.[30] The task of compensating for inadequate secondary education among minority populations prompted prominent HBCUs such as Hampton and Tuskegee to inaugurate pre-engineering programs, in which students began training locally but completed their degrees at historically white institutions with superior engineering programs.[31] Such programs provided important models that this study will explore, but many African-Americans attended schools with fewer financial resources and faced severely limited opportunities for advanced technical training.[32] To capture this more representative type of experience I will examine engineering curricula at two Chicago schools originally established to serve working-class populations and attended in this period largely by black students and white students, respectively: Kennedy-King College (formerly Wilson Junior College) and the Illinois Institute of Technology. While each had its origins in trades training, IIT shifted in this era to a more professionalized curriculum than did Kennedy-King. I will explore this pedagogical divergence along lines of race as a set of social and political choices.

Here my study moves from land-grant to urban schools. Whether public or private, urban colleges and universities in these years often saw themselves as having a prominent role in the betterment of a "troubled society."[33] IIT, having grown from the Lewis and Armour Institutes, saw its legacy as an inner-city school as demanding participation in urban renewal and growth in Chicago, but the benefits of this progressive attitude for minority engineering remain to be explored.[34] Like Columbia University and the University of Pennsylvania, IIT appears to have made a commitment to liberal social agendas in the 1960s that involved limited address of student demographics or curricular content.

In scrutinizing choices made by Chicago's educational institutions in this period we must consider broad patterns of private and public funding. IIT is a private school that prior to the 1960s derived much of its revenue from evening classes attended by working-class Chicagoans. When the state-funded University of Illinois at Chicago greatly expanded its nearby campus and began to offer lower-priced degree courses, IIT lost the bulk of its enrollment base. IIT administrators responded by shifting the school's mission to one based on a more selective day-student enrollment and on high-level research, but in so doing, IIT became less accessible to students emerging from the city's disadvantaged public high schools—many of whom were African-American. To understand this shift we must consider IIT's dependence on local interests that controlled economic resources in the state; neither industrial nor government funding sources moved to preserve IIT's neighborhood identity, perhaps preferring (realistically or otherwise) that the University of Illinois's new Circle Campus (UIC) fulfill that function. In any case, IIT ceased to serve the South Side in the way its founders may have envisioned. It is highly significant that IIT established an early "Minorities in Engineering" program in 1974 that brought relatively large numbers of black and Hispanic students to the school, and with exceptionally high retention rates, and it would be inappropriate to ascribe the school's departure from its traditional service role to racialist thinking in any direct way. And yet, its curricular design (like its razing of a large black neighborhood to create its modernist campus) had consequences for black Chicagoans that deserve historical analysis.

A similarly complex cultural and political narrative, involving many players beyond individual institutional administrators, may be written about the public educational options available to Chicago's aspiring engineers in the 1960s and the 1970s. While the UIC did offer an affordable education in many technical fields, black Chicagoans seeking post-secondary education turned in greater numbers to the city's junior and city colleges, a tier of urban education about which almost no history has been written. Their choice to do so may have been contingent on "pipeline" issues (which attribute students' post-secondary disadvantages to weak secondary training), but pipeline explanations generally fail to locate larger causes of such societal inequities or, as I hope to do, explain in positive terms patterns of post-secondary education. Kennedy-King College stood out in the Chicago City College system as an institution particularly committed to the predominantly black neighborhoods of South Side Chicago. For the school's directors during the 1960s and the 1970s, such a commitment meant the creation of a "comprehensive" curriculum geared toward the diverse abilities and ambitions of that community, rather than development of the school's original industrial emphasis.[35] I will examine tensions surrounding this choice of mission and their eventual resolution in a system-wide reorganization and the creation of specialized city college branches for engineering. As at IIT, issues of school identity—established along lines of locality or race—configured educational planning in Chicago's city colleges, and through these cases we will see how such factors shaped the seemingly value-neutral pursuit of "technical excellence."[36]

1980s and 1990s

In the final decades of the twentieth century, two trends brought new pressures to university engineering programs. Instruction in what was already a cost-intensive set of disciplines became even more expensive as the pace at which technical knowledge and equipment moved into obsolescence quickened.[37] This era also saw a conservative retrenchment in American politics in the form of anti-Affirmative Action initiatives and general reductions in federal involvement in higher education.[38] I focus for this period on the teaching of computing and related "high-tech" subjects at Prairie View University and at Texas Agricultural and Mechanical (A&M) University.

For many years, engineering facilities at these two Texas land-grant institutions were of drastically different quality, the historically black Prairie View operating at a tremendous deficit.[39] Texas A&M had been well funded since its establishment as the state's first public institution of higher learning and it has long attracted many of the state's wealthiest families as an alternative to Ivy League schools. Today Texas A&M holds the designation not just of land-grant, but also of sea-grant and space-grant university, consistently ranking among the top ten U.S. universities in dollar-value of its research.[40] An influx of funding in the late 1980s allowed Prairie View to enhance its operations: among its initiatives was an honors program that included new colleges of Applied Science and Engineering Technology and of Engineering and Architecture.[41] Prairie View had, even in the early 1980s, been a leading producer of black technical graduates, and these programs were intended to halt the "brain drain" of students to white universities.[42] I will examine here controversies surrounding decisions to enhance engineering programs at HBCUs rather than expend resources on established white schools. Interestingly, these decisions were resisted by both liberal elements that decried the perpetuation of racially distinct educational settings and conservative forces concerned about the "waste" of duplicating facilities.

The two sides were clearly invoking differing conceptions of the public good. All solutions offered for racial inequities carry broad political implications, and each projects a particular degree or rate of systemic change—a spectrum of radicality to which this study will be sensitive. In this period, an enduring liberal-moderate coalition in Texas moved into an era of decreased public influence; social services for minorities and historically disadvantaged groups diminished.[43] Notably, many new public policies were helpful to big business, and higher education in technical fields had a role to play in these developments. Service to industry through the provision of graduates and research has shaped American university engineering programs since their inception in the mid 1800s; here I will investigate the mutual influence of industrial expectations and academic approaches to diversification in and beyond Texas.[44] For example, despite overarching shifts to conservative social policies, the 1980s and the 1990s saw an increase in corporate sponsorship of minority mentoring and remediation programs in higher education. While such sponsorship clearly has benefits for those students to whom it is made available, this type of program may offer a less comprehensive agenda for social reform in engineering and the sciences than would, say, Affirmative Action, and I will explore the political implications of such corporate activity in this final portion of my study.[45]

Methodology

As I examine the ways in which university administrators and faculty approached engineering education, I will measure instructors' intentions against the perceptions of engineering students as reported in oral history interviews, a juxtaposition that will highlight discontinuities between the stated and actual natures of technical programs—potentially significant disparities, as we learn from historians of education.[46] Further, the ways in which students identity themselves—by ethnicity, gender, economic background, future occupation, or none of the above—may differ drastically from classifications applied by administrators, instructors, or educational analysts, and in any study of social groups such variations are crucial, as well.[47] The experiences and priorities of white and non-white individuals will both be central to this research; in choosing subjects for oral history interviews, I will seek faculty and graduates of all eight schools under study. I will also interview engineers who are prominent in educational reform and represent the agendas of professional engineering associations, government agencies, and other controlling entities in the education of American engineers.

In addition, there will no doubt be considerable variation in how interviewees assess the character of their education. In some cases discrimination will not be a relevant measure, while in others it may surface early in conversation.[48] To suggest that racial biases or other social factors shaped an engineer's career may be to challenge the terms in which that engineer sees his or her achievements and experiences. One goal of this project is to explore how the historian's politicized view of science and engineering more generally meshes with or contravenes the outlooks of practicing technical professionals. If studies such as this one are to be suggestive of changes in educational methods, we must test the validity of STS premises in this way. The two views need not be congruent in order for STS to yield applicable findings, but differences between the perspectives must be clear.

In future projects, I will carry my study of race and engineering to the technical workplace. The idea that one can locate systems of social privilege by studying the epistemological features of engineering is relatively new to historians, but one that I believe holds great promise. While the successes of individual schools and individual students are undeniable, the differential rates at which minorities and non-minorities are entering engineering fields may reflect systemic obstacles: in government and institutional policies, but possibly also in the practices of science and engineering themselves. Are features of these disciplines inherently conservative—either in their reification of existing industrial hierarchies of skill, or in their very definitions of technical proficiency? Or, do the rigor and cultural influence of science and engineering lend them special progressive potential? Such considerations are the overarching concerns of this project.

Notes

1. Among the most comprehensive recent studies are Willie Pearson Jr. and H. Kenneth Bechtel, *Blacks, Science, and American Education* (Rutgers University Press, 1988); Michael T. Nettles and Laura W. Perna, *The African American Education Data Book* (Frederick D. Patterson Research Institute, 1997); and Antoine M. Garibaldi, "Four Decades of Progress," *Journal of Negro Education* 66 (1997): 105–120. On the history of race-sensitive college admissions in general, see William G. Bowen and Derek Bok, *The Shape of the River: Long-Term Consequences of Considering Race in College and University Admissions* (Princeton University Press, 1998).

2. A few studies reflect statistical change over time. See Willie Pearson Jr. and Larue C. Pearson, "Baccalaureate Origins of Black American Scientists: A Cohort Analysis," *Journal of Negro Education* 54 (1985), p. 24; Garibaldi, "Four Decades of Progress." The only synthetic history on this topic is David E. Wharton, *A Struggle Worthy of Note: The Engineering and Technological Education of Black Americans* (Greenwood, 1992). Clarence Williams's book *Technology and the Dream: Reflections on the Black Experience at MIT* (MIT Press, 2001) supplies the only truly comprehensive portrait of black experiences within a technical institution, however atypical MIT may be.

3. For overviews of approaches that will inform this study, see Henrietta Kuklick and Robert Kohler, "Introduction—Science in the Field," *Osiris* 11 (1996): 1–11; Jan Golinski, *Making Natural Knowledge: Constructivism and the History of Science* (Cambridge University Press, 1998).

4. For one of very few articles to focus on equipment used in engineering education (in this case, computers), see Randolph Hawkins and Arthur E. Paris, "Computer Literacy and Computer Use among College Students: Differences in Black and White," *Journal of Negro Education* 66 (1997): 147–158.

5. Elaine Seymour and Nancy M. Hewitt, *Talking about Leaving* (Westview, 1997); Amy Slaton, *Reinforced Concrete and the Modernization of American Building* (Johns Hopkins University Press, 2001), chapter 1.

6. Because race has not figured prominently in STS research, gender-centered works will be valuable models for this study. See Judy Wajcman, *Feminism Confronts Technology* (Pennsylvania

State University Press, 1991); Margaret Rossiter, *Women Scientists in America: Before Affirmative Action, 1940–1972* (Johns Hopkins University Press, 1995); Nina Lerman, From "Useful Knowledge" to "Habits of Industry": Gender, Race, and Class in 19th-Century Technical Education, Ph.D. dissertation, University of Pennsylvania, 1993; Judith McGaw, "Women and the History of Technology," *Signs* 74 (1982): 798–828; Mary Frank Fox, "Women and Scientific Careers," in *Handbook of Technology Studies*, ed. S. Jasanoff et al. (Sage, 1995); Sandra L. Hanson, *Lost Talent: Women in the Sciences* (Temple University Press, 1996).

7. Golinski, *Making Natural Knowledge*. See also Robert Kohler, *Lords of the Flies: Drosophilia Genetics and the Experimental Life* (University of Chicago Press, 1994); Kuklick and Kohler, "Introduction."

8. For overviews of approaches that will inform this study, see Kuklick and Kohler, "Introduction"; Golinski, *Making Natural Knowledge*. Science studies and related fields have not looked extensively at issues of race; the following are important exceptions: Sandra Harding, ed., *The "Racial" Economy of Science* (Indiana University Press, 1993; Lerman, "From 'Useful Knowledge' to 'Habits of Industry.'"

9. William H. Martin, "The Land-Grant Function of the Negro Public College," *Journal of Negro Education* 31 (1962): 396–403.

10. The first Ph.D. in civil engineering to be given to a black student was awarded in 1939, the first Ph.D. in chemical engineering, in 1943. Life sciences granted doctorates to African-Americans somewhat earlier. See Caldwell Titcomb, "The Earliest Ph.D. Awards to Blacks in the Natural Sciences," *Journal of Blacks in Higher Education* (1997): 92–99.

11. Warmoth T. Gibbs, "Engineering Education in Negro Land-Grant Colleges," *Journal of Negro Education* 21 (1952): 546–550; Susan T. Hill, *The Traditionally Black Institutions of Higher Education, 1860–1982* (National Center for Education Statistics, U.S. Department of Education, 1985); James D. Anderson, *The Education of Blacks in the South, 1869–1935* (University of North Carolina Press, 1988); Robert L. Jenkins, "The Black Land-Grant Colleges in Their Formative Years, 1890–1920," *Agricultural History* 65 (1991): 63–72. See also Nina Lerman's essay in this volume.

12. John Fleming, *The Lengthening Shadow of Slavery: A Historical Justification for Affirmative Action for Blacks in Higher Education* (Institute for the Study of Education Policy, Harvard University, 1976), p. 58. On contemporary medical education, see Rufus E. Clement, "The Impact of the War upon Negro Graduate and Professional Schools," *Journal of Negro Education* 11 (1942), p. 371; Wharton, *A Struggle Worthy of Note*, pp. 61–74.

13. Fred McQuiston, *Graduate Instruction for Negroes* (George Peabody Institute for Teachers, 1939); Federal Security Agency, *National Survey of the Higher Education of Negroes* (Government Printing Office, 1942).

14. Huber O. Croft, *A Brief History of the College of Engineering, University of Missouri—Columbia* (University of Missouri, 1968), pp. 28–29; Wharton, *A Struggle Worthy of Note*, p. 76; Mendell P. Weinbach, *Engineering at the University of Missouri, 1850–1940* (Engineering Foundation, University of Missouri, 1941); Frank F. Stephen, *History of the University of Missouri* (University of Missouri, 1962).

15. George E. Simpson and J. Milton Yinger, *Racial and Cultural Minorities: An Analysis of Prejudice and Discrimination* (Harper & Row, 1972).

16. William S. Savage, *The History of Lincoln University* (Lincoln University, 1939); Alpert P. Marshall, *Soldier's Dream: A Centennial History of Lincoln University of Missouri* (Lincoln University, 1966); Lewis K. Downing, "The Negro in the Professions of Engineering and Architecture," *Journal of Negro Education* 4 (1935): 60–70; Walter C. Eells, "Surveys of Higher Education for Negroes," *Journal of Negro Education* 5 (1936): 245–251; Robert G. Cotton, "Collegiate Technical Education for Negroes in Missouri," *Journal of Negro Education* 15 (1946), p. 174.

17. Willie Pearson Jr. and Alan Fechter, eds., *Who Will Do Science?* (Johns Hopkins University Press, 1994), p. xiii.

18. Cotton, "Collegiate Technical Education for Negroes in Missouri," p. 172; E. Worthington Waters, "Problem of Rural Negro High School Seniors on the Eastern Shore of Maryland: A Consideration for Guidance," *Journal of Negro Education* 22 (1953), p. 115.

19. D. A. Hollinger, "Inquiry and Uplift: Late Nineteenth-Century American Academics and the Moral Efficiency of Scientific Practice," in *The Authority of Experts*, ed. T. Haskell (Indiana University Press, 1984); Slaton, *Reinforced Concrete*.

20. Cheryl B. Leggon and Shirley M. Malcom, "Human Resources in Science and Engineering: Policy Implications," in *Who Will Do Science*, ed. Pearson and Fechter, pp. 145–148.

21. In 1855, 0.65 percent of all engineering graduates were black; in 1970, 0.67 percent. See Philip Carey, "Engineering Education and the Black Community: A Case for Concern," *Journal of Negro Education* 46 (1977), p. 43.

22. Gladyce H. Bradley, "Negro Higher and Professional Education in Maryland," *Journal of Negro Education* 17 (1948), p. 303; G. James Fleming, "Racial Integration in Education in Maryland," *Journal of Negro Education* 25 (1956); Governor's Commission to Study the Problem of the Expansion of the University of Maryland, *A Plan for the University of Maryland* (University of Maryland, 1960); George H. Callcott, *A History of the University of Maryland* (Maryland Historical Society, 1966), pp. 346, 394; Ernest O. Campbell, "Negroes, Education and the Southern States," *Social Forces* 47 (1969): 253–265.

23. Callcott, *History of the University of Maryland*, pp. 357–358.

24. Ibid., p. 347.

25. Ibid., p. 351.

26. Ibid., p. 352.

27. The black normal schools also helped white public-school teachers avoid service in black communities; see Eldon L. Johnson, "Misconceptions about the Early Land-Grant Colleges," *Journal of Higher Education* (1981); Anderson, *Education of Blacks in the South*; Marsha L. Matyas and Shirley M. Malcom, *Investing in Human Potential: Science and Engineering at the Crossroads* (American Association for the Advancement of Science, 1991). UMD's decision is representative of a larger historical pattern: in 1994, 54 percent of all doctorates awarded by American HBCUs were in the field of education; see Robert B. Slater, "Rating the Science Departments at Black Colleges and Universities," *Journal of Blacks in Higher Education* 4 (1994), p. 90.

28. Wilson H. Elkins, *A Decade of Progress and Promise, 1954–1964: A Report to the Board of Regents of the University of Maryland* (University of Maryland, 1964), p. 28.

29. Wilbert Randolph Wilson, *An Historical Analysis of Events and Issues Which Have Led to the Growth and Development of the University of Maryland, Eastern Shore, from 1886 to 1975* (George Washington University, 1976); Carl S. Person, *Revitalization of an Historically Black College: A Maryland Eastern Shore Case* (House of Representatives, 1998).

30. Leggon and Malcom, "Human Resources in Science and Engineering," 147; Frederick D. Patterson, "Duplication of Facilities and Resources of Negro Church-Related Colleges," *Journal of Negro Education* 29 (1960); William M. Boyd II, *Desegregating America's Colleges: A Nationwide Survey of Black Students, 1972–1973* (Praeger, 1974); National Science Foundation, *Characteristics of the National Sample of Scientists and Engineers* (National Science Foundation, 1974 and 1982); Wharton, *A Struggle Worthy of Note*; and Bowen and Bok, *Shape of the River*.

31. William Trent and John Hill, "The Contributions of Historically Black Colleges and Universities to the Production of African American Scientists and Engineers," in *Who Will Do Science*, ed. Pearson and Fechter.

32. Abram J. Jaffe, Walter Adams, and Sandra G. Meyers, *Negro Higher Education in the 1960's* (Praeger, 1968), pp. 134–138.

33. Kermit C. Parsons, "A Truce in the War between Universities and Cities: A Prologue to the Study of City-University Renewal," *Journal of Higher Education* 34 (1963): 16–28.

34. Michael Kennedy, "AFGRAD: An Experiment in Purposive Education: The First Four Years of the African Graduate Fellowship Program," *Journal of Higher Education* 38 (1967): 500–506; Irene Macauley, *The Heritage of Illinois Institute of Technology* (Illinois Institute of Technology, 1978).

35. John F. Grede, "Collective Comprehensiveness: A Proposal for a Big City Community College," *Journal of Higher Education* 41 (1970): 179–194; Peter A. Remus, *A Historical Study of the Department of Architecture and Engineering at Kennedy-King College* (Chicago State University, 1972).

36. The role of junior or community colleges in the technical training of minorities is a vital historical topic: the proportion of college-going African-Americans attending two-year rather than four-year schools remains at close to 50 percent (Garibaldi, "Four Decades of Progress," p. 114).

37. Trent and Hill, " Contributions of Historically Black Colleges and Universities"; Richard Seltzer et al., "Multiculturalism, Race, and Education," *Journal of Negro Education* 64 (1996): 124–140.

38. Russell L. Riley, *The Presidency and Politics of Racial Inequality: Nation-Keeping from 1831 to 1965* (Columbia University Press, 1999). In 1991, the total amount of federal science and engineering grants to all 100 existing HBCUs was less than that to *each* of Johns Hopkins, Stanford, and MIT, suggesting the extent to which black schools suffered from secondary technical status (Slater, "Rating the Science Departments," p. 90).

39. Alton Hornsby Jr., "The 'Colored Branch University' Issue in Texas: Prelude to Sweatt vs. Painter," *Journal of Negro History* 61 (1976): 51–60; Downing, "The Negro in the Professions of Engineering and Architecture," 90.

40. Henry C. Dethloff, *A Centennial History of Texas A&M University* (Texas A&M University Press, 1975).

41. Elaine P. Adams, "Benjamin Banneker Honors College: Gateway to Scientific and Technical Doctorates," *Journal of Negro Education* 59 (1990): 449–462.

42. Gail Thomas, "Black Participation and Performance in High School Science," in Pearson and Bechtel, *Blacks, Science, and American Education*, p. 72; Adams, "Benjamin Banneker Honors College."

43. S. H. Murdock, *The Population of Texas: Historical Patterns and Future Trends Affecting Higher Education* (Texas A&M University Press, 1989); Chandler Davidson, *Race and Class in Texas Politics* (Princeton University Press, 1990).

44. David Noble, *America by Design: Science, Technology, and the Rise of Corporate Capitalism* (Oxford University Press, 1977); Bruce Seely, "Research, Education, and Science in American Engineering Colleges: 1900–1960," *Technology and Culture* 34 (1993): 344–386.

45. Edmund W. Gordon, *A Descriptive Analysis of Programs and Trends in Engineering Education for Ethnic Minority Students* (National Action Council for Minorities in Engineering, 1986); Leggon and Malcom, "Human Resources in Science and Engineering."

46. An important guide for this investigation of classroom experiences will be Larry Cuban's *How Teachers Taught: Constancy and Change in the American Classroom* (Teachers College Press, 1993).

47. In *Technology and the Dream* Clarence Williams captures some of the disparities in such classification in his accounts of black faculty, staff, and students at MIT, an elite institution that projected a variety of identities on its nonwhite members.

48. Cuban, *How Teachers Taught*; Valerie Raleigh Yow, *Recording Oral History: A Practical Guide for Social Scientists* (Sage, 1994).

Museums and the Interpretation of African-American History

Lonnie Bunch

In the late 1970s, African-American veterans, especially members of the all-black fighter squadrons known as the Tuskegee Airmen, claimed that the National Air and Space Museum intentionally underplayed the important contributions of black aviators in World War II.[1] This controversy about race led a few members of Congress, most notably Senator Edward Kennedy, to inquire about the presentation of African-American history at the Smithsonian Institution. One of the ways the museum responded to these public and congressional inquiries was to ask several African-American staff members to allow their likenesses to grace mannequins that would be placed in the museum in order to increase the "black presence" in its exhibits. But these figures were positioned in airplanes and settings so high, or far removed from view, that the only way the public could see this increased presence was by scanning the outer reaches of the museum, using binoculars purchased in the gift shop.

In today's museums, African-American history clearly is no longer on the fringes of the stage. For those who study or are interested in the African-American past, there have been many imaginative exhibitions that have stretched the interpretive parameters and challenged the tenor and color of the museum profession's historical presentations. The past decade has been a period of growth, excitement and possibility. Museums as diverse as the National Museum of American History, the Museum of the Confederacy, the Oakland Museum, and the Henry Ford Museum have wrestled creatively with African-American subject matter. Of even greater import and instruction is the array of smaller institutions, including the Please Touch Museum in Philadelphia, the Geneva (New York) Historical Society, and the Northern California Center for African-American History and Culture, that have sought to give local meaning to the issues of race in America.

As a consequence of these activities, the public has experienced exhibitions that explored slavery, African-American migration from the South to the North, craft skill, invention, Africanisms in American culture, representations of race in American art, urbanization and community development, and issues at the intersection of race and gender. This research and exhibition of African-American life in museums has contributed a vibrancy and relevance that has invigorated many of the nation's cultural institutions, and sparked useful collaborations between museums and their local communities.

While there have been great changes in whom and what museums interpret, it is much too soon to be satisfied with the profession's efforts in exploring African-American culture. Often the rhetoric of change fails to match the realities of everyday life in museums. My major concern is that museums are too often crafting exhibitions that simply say "African-Americans were here too," rather than examining the complexities, interactions and difficulties of race in America. In essence, much of what museums create today is better suited to the world of 40 years ago when, as the novelist Ralph Ellison suggested, blacks were invisible men and women, and whites needed to be reminded that African-American history and culture mattered. Presentations for the next century need to reflect better the clashes, compromises, broken alliances, failed expectations and contested terrain that shape the perspectives of today's audiences.

Yet there is no denying that more historical institutions have embraced the opportunity to explore the African-American past than ever before. Much of this new openness stems from broader cultural and societal concerns that have influenced and shaped the world of museum professionals. The struggles of the Civil Rights and Black Power movements in the years after World War II seared the issues of race and justice into the nation's consciousness and made the African-American experience more immediate, more visible, and more important to America's museums.

Despite an uneasy marriage between museums and academic scholarship, the relationship has provided museum curators and directors with new tools, new challenges, and new possibilities. Especially important was the impact of the "New Social History," with its clarion call to explore America's diverse past "from the bottom up." As a generation of scholars trained in this different approach to historical analysis entered the museum field in the 1960s and the 1970s, they brought new questions and interests that stimulated research, collection policies, and exhibitions in African-American history from the perspective of that community.

The shifting demographics of many communities served by museums also ensured that, in the words of the singer Sam Cooke, "a change is gonna come." Cooke's song poignantly expressed the hope for improved relations between blacks and whites:

I was born by the river in a little tent
And just like the river, I've been running ever since
It's been a long time coming
But I know a change is gonna come

It's been too hard living, but I'm afraid to die
I don't know what's up there beyond the sky
It's been a long time coming
But I know a change is gonna come

I go to the movie, and I go downtown
Somebody keep telling me "Don't hang around"
It's been a long time coming
But I know a change is gonna come

Then I go to my brother and I say, "Brother help me please"
But he winds up knocking me back down on my knees

There've been times that I've thought I couldn't last for long
But now I think I'm able to carry on
It's been a long time coming
But I know a change is gonna come.

As cities and suburbs were reshaped by black in-migration and white flight, many institutions realized that they had to become more responsive to the needs of their African-American neighbors, especially if they were to remain lively and interesting places. And in many of those cities, the new African-American political leadership demanded greater political accountability: If museums were to utilize public funding, then their programs, exhibitions, and collections must reflect the diversity of the region.

An important and often unacknowledged contributor to this increased African-American presence was the National Endowment for the Humanities. At a time when the Endowments are under attack for being either superfluous or partisan, it is important to remember that for nearly three decades the NEH has been a pivotal funding agent for America's museums of history. By awarding support to institutions that explored and exhibited the black past, the NEH both helped legitimize and encourage the place of African-American history in museums. Ultimately, cultural institutions began to embrace the African-American past once they realized what the university community has known for decades: that African-American history is a dynamic, stimulating field with broad popular appeal that illuminates many aspects of American history.

Despite this decade or so of substantial progress and change, there is a need to move the presentation of African-American history in America's museums to a higher interpretive plane—to a level that embraces a more holistic and diverse view of the African-American experience, that recognizes the need for new paradigms and alternative structures to shape both the products and process of exploring the black past in museums, and that makes the African-American past usable for all Americans. What follows are several suggestions, and a exemplary case, that I think will enhance the ability of public historians to explore and to make accessible the African-American past.

It is necessary first to resist monolithic portrayals of African-American life. When one reads African-American literature, whether it is the urban poetry of Langston Hughes, the rich depictions of racial joys and sorrows in the work of Maya Angelou, or the musings of Terry McMillan, one is struck by the richness of the mosaic of African-American life. In this literature, one is introduced to a black world that abounds with differences based on class, religion, gender, color, political perspective, and education. Yet far too often most museum exhibitions fail to convey this rich diversity or to explore the meaning of these differences for the audience.

Usually the prism through which an exhibition explores the black community is that of the middle class. While the black middle class is central to understanding much about black life and aspirations, delineating only that history obscures the full range of African-American experiences. Because the middle class has traditionally been but a small segment of the African-American community, it is essential that institutions expand both the subjects and the perspectives of their public presentations. By resisting this tendency to uniform descriptions of a people, museums can help visitors better understand the conflicts, negotiations, and shifting coalitions that have historically composed most black communities. By choosing for exhibition such subjects as labor practices, industrial education, gender roles, burial traditions, craft skills, and storefront religions, cultural institutions are more likely to provide a richer, more complex, and more complete lens into the African-American past.

The second thing we need to do is transcend an idealized view of the past. Carter G. Woodson, an African-American scholar who is best remembered for the creation of Black History Month, was once asked why he became a historian. He replied: "The only reason that I do history is to make America better."[2] If museums are to follow the words of Woodson, it is important that the profession avoid romanticizing the past, especially African-American history. Too often, exhibitions about race exude a rosy glow. The African-American community is characterized as comprising upwardly mobile heroes, to whom racism and discrimination were simply obstacles to overcome. While that scenario did occur, it was the exception and not the rule during much of America's history.

What is wanted is a commitment to explore the full range of African-American experiences, including the difficult, controversial and ambiguous episodes. It is essential that the harsh realities of black life are seen side by side with achievements and victorious struggles. The Valentine Museum and the Museum of the Confederacy, both in Richmond, Virginia, provide good examples of the presentation of slavery and race relations in the urban South. Visitors are offered a richly variegated portrait that is replete with great joy and great sorrow, and all that is in between—in essence, the stuff of real history. Few institutions, however, are as willing to exhibit the difficult aspects of the African-American past. Fewer still wrestle effectively with issues of violence, riots, lynching, with technological unemployment, and the devastating effects of generations of poverty and discrimination. I am not arguing that museums focus only on the harsh or the unpleasant, or depict African-Americans as simple victims of history. But I do feel that museums must do a better job of mirroring the complexities and ambiguities of black life.

The third thing we need is a new set of paradigms, in order to deal effectively with a more subtle and interesting history. In many cultural institutions, presentations of the African-American past are shoe-horned into the usual museum structures, interpretive devices, and modes of display—crafted by a traditionally trained staff that is usually less than diverse. While this formula has produced important exhibitions, it is now appropriate that museums rethink and expand these traditional paradigms in order to convey the full complexity and richness of black life. For example, some museums have already begun to shift their involvement with African-American communities into a new model that recognizes the importance of developing long-term and mutually reciprocal relationships. Movement in this direction is crucial because it is not easy to tell an evocative and nuanced history without the knowledge and respect of the living African-American community. This paradigm of shared responsibility recognizes the difficulties and the benefits of community influence in the development of exhibitions, programs, and collections. This is not to suggest that curators abrogate their scholarly and professional obligations, but rather that they embrace a new way of thinking that accepts the notion that strong exhibitions often grow out of balancing the tensions between community memory and academic history.

By themselves, however, new community-museum paradigms are not enough. Interpreting African-American history—in fact, exploring all of the American past—can benefit greatly from the creation of an internal environment that encourages innovation, creativity, and respect for differing cultural perspectives. Interpreting African-American history in new ways could mean an explicitly interdisciplinary approach to the past that enriches the traditional processes and visions of historians by adding folklorists, artists, ethno-musicologists, and archaeologists. Though museums have long expressed interest in an interdisciplinary approach, the actual implementation of this partnership has been too sporadic and uneven. Even more important is the willingness of the cultural entities to stretch the parameters of traditional museum interpretation. Experimenting with content, manipulating the role and use of objects, marrying new technologies with the established interpretive devices, and expanding visual opportunities can lead to exhibitions, such as Fred Wilson's "Mining the Museum" at the Maryland Historical Society in 1993, that provide insights, engagement, learning, and real understanding about the importance and the centrality of race in American culture.

Finally, what we need is to discover a "new integration." In 1896, the United States Supreme Court heard a case, *Plessy v. Ferguson*, pertaining to a Louisiana law that required racially segregated railroad facilities. The Court ruled that segregation of railroad accommodations was constitutional as long as "separate but equal" facilities were available. This decision ensured that segregation was the law of the land throughout much of the twentieth century. While this judgment was ultimately overturned in 1954, thus legally ending segregation in America, the doctrine of "separate but equal" is very much alive in many cultural institutions. Far too frequently, African-American history is segregated from the "other" history that museums explore. African-American history is often interpreted by staff, and seen by visitors, as an interesting and occasionally instructive episode that has limited meaning for most non-African-Americans. This is not to demean the efforts of museums. Clearly, it is important that museums continue to craft exhibitions that explore specific aspects of black culture and history. There is always a need for exhibits that explore the history of the African-American in Cleveland, or the African-American image in advertising. But these depictions only convey part, albeit a significant part, of the story of the African-American past.

What is needed is a new synthesis, a "new integration" that encourages visitors to see that exploring issues of race generally, and African-American history specifically, is essential to their understanding of American history. It is useful that museums convey both the importance of race in American history, and the fact that race does not simply mean "people of color." At a time when racial concerns influence our perceptions on everything from the current political debates, to the O. J. Simpson trial, to the state of popular music, or access to the Internet, museums can perform a valuable service by demonstrating how issues of race have touched, shaped, and informed—historically— the experience of all Americans. The language of billboards mounted throughout Los Angeles by the California Afro American Museum in the late 1980s argued this point with the message "Our history makes American history complete."

The 1932 transcontinental airplane flight of Herman Banning and Thomas Allen offers us an example of African-American history that informs and extends our understanding of America's experience with this new technology. In the first instance, the cast of characters is interestingly enlarged. Besides Charles Lindbergh, Amelia Earhart, and other celebrated idols of the golden age of American aviation, we could now include William Powell, Hubert Julian, and Bessie Coleman, as well as Banning and Allen—black aviators who flew despite "an almost universal prejudice against Negroes" among white pilots. Lindbergh and Earhart enjoyed financial backing and sophisticated technological support systems; Banning and Allen's flight illuminates the dark side of our aviation history.

One of the pioneers of black aviation and the most experienced pilot of his race, Banning had advantages that other blacks did not. He grew up in a middle-class family in Ames, Iowa, and attended the state university there for three and a half years. Banning learned to fly from a World War I veteran and by 1927 was one of only a handful of blacks who possessed a commercial pilot's license. On the basis of those skills, he established a number of "aero clubs" in Los Angeles that stimulated air-mindedness in the black community there. Thomas Allen, like many young Americans of that era, caught aviation fever and worked as a handyman and mechanic's helper in airfields throughout the Southwest in order to earn money for flying lessons. He drifted into Los Angeles, contacted one of the aero clubs, and so met Banning, who was looking for a co-pilot. With $25 in their pockets, they took off from Los Angeles airport on the afternoon of September 21, 1932, in a four-year-old plane with a 14-year-old engine, leaving behind a "crowd" of four people who had come to wish them well.

It tells us something important that their flight was partly motivated by a rumor floating around the airport about a prize of $1,000 that would go to the first blacks to complete a transcontinental flight successfully. During the depths of the Great Depression, when aero clubs languished for want of money, the prospect of winning $1,000 was reason enough to try the impossible. But Banning and Allen were also interested in accomplishing a "race first." A cross country flight was "just something which had never been done before by members of our race," they explained, "and we thought we would do it."[3] The aviation world of the 1920s and the 1930s was full of contests, prize awards, and newspaper coverage; everyone—blacks and whites alike—wanted to believe in the transforming power of flight.[4] New technologies would be the solution to the nation's economic problems, to the social ills of crowded cities, and for African-Americans in particular, to the evils of racism. Since flying involved skills beyond the grasp of most Americans, even as it captured their imaginations, the sight of blacks in command of these wondrous machines would surely shatter the myth of their technological inferiority.

Aviation of this period was also emblematic of a new, urbanized, industrialized society, and leaders of the black community were concerned to have a place in it. Race pride was one way to that end. Marcus Garvey captivated the black masses by proclaiming "up, up you mighty race."[5] Abuzz with its jazz clubs and cabarets, poets and playwrights, Harlem established itself as the "capitol of the Negro world." Black people, according to Alain Locke, would no longer wait for equality; the "New Negro" became more assertive, militant and unwilling to accept an inferior place in society:

"Negroes want to be seen in the same pattern, only in a different shade."[6] In this context, the success of one black person represented a success shared by all, and Banning and Allen knew they carried such hopes.

The African-American press came to call them "black eagles," and Banning and Allen called their airplane "Eagle Rock." But they called themselves "flying hobos," because they depended upon the good will of the people they met along the way for gasoline, spare parts, food, drink, and a bed for the night. Since cool weather had already set in, they took a southern route, stopping in Tucson, El Paso, Oklahoma City, Wichita Falls, St. Louis, Pittsburgh, and Trenton. At each stop, Banning and Allen sought out the black section of town, where the "birdmen" ate and lodged free. Sometimes the townspeople even washed their airplane and filled its fuel tanks. All along the way, the black community turned out to welcome them as celebrities. From El Paso to Terre Haute, they were treated not as hobos but as black heroes struggling against the odds.[7]

And what gives their story extra dimension is that they received a surprising amount of support from white people, even in places where they might have expected the opposite. In Wichita Falls, Texas, for instance, rather than seeing only black faces when they landed, they were welcomed by the whole community as "the first colored fliers to land here." The local press covered the event, and the townspeople provided free food and transportation. In St. Louis, when they realized they had serious engine problems, a local white mechanic agreed to repair the motor for little or no cost. When their money ran out in Pittsburgh, the *Courier* arranged for the local Democratic party organization to fund the remaining leg of the flight, providing the pilots dumped campaign literature along the route.[8] And when they finally landed at Roosevelt Field on Long Island, three weeks after their departure from Los Angeles, a real crowd greeted them.

Indeed, Manning and Allen became the toast of New York. City officials presented them with medals, there were speeches and press interviews. A banquet crowd honored them at the Marlboro Restaurant, one of Harlem's finest establishments, while other restaurants and hotels scrambled to give them free room and board. And for nine weeks, the Pittsburgh *Courier* published a column they wrote called "Coast to Coast via the Aerial Highways." Each column, which began "Good morning fellow aviator," read like an adventure story, as if readers were actually in the cockpit of the plane.

The popular response to Lindbergh's flight cast him as a solitary hero. He always preferred to celebrate his machine's modern technology, but the rhetoric surrounding his feat reached back in time for notions of rugged individualism and the hardy spirit of nineteenth-century pioneers.[9] By contrast, America's black community focused on the collective importance of the Manning and Allen flight, especially as it bore on the race's future. "They have bridged the barrier of color," one newspaper proclaimed, and in the process became examples for future generations of young people. Inevitably, they were described as "suntanned editions of the Lindy of yesteryear," but such rhetorical flourishes aside, the language of the black press emphasized the "New Negro's" concern for role models and race heroes. Many blacks felt the lack of positive images had limited the horizons of black youth. The time had come, according to the *Negro World*, "for the Negro to forget and cast behind him his hero worship and adoration of other races, and to start out immediately to create and emulate heroes of his own."[10] Banning and Allen, "two modest and unassuming chaps" who had just set an aviation record, filled those needs.

Their flight also captured the imagination of the black community because of its pioneering character, though here again, African-Americans used the term differently than their white counterparts did when writing about Lindbergh. In this case, pioneering meant competing with whites in a new technology as a way of achieving racial justice. Aviation presented blacks with the opportunity to fly as fast and as far and as high as white aviators. So, the flight's greatest significance was as a giant step toward equality. How could white America not see that black equality in the air should lead to black equality on the ground? Charles Johnson, editor of *Opportunity*, the journal of the Urban League, wrote that "aviation has tremendous and dramatic possibilities to help the race in its fight for world wide respect." W. E. B. Du Bois put it more poetically when he said, the black race "soars upward, on the wings of an aeroplane."[11]

The transcontinental flight of Banning and Allen tells us more than a story of personal courage—though there was that. It throws light on the way Americans of different class and color responded to the possibilities they thought inherent in technical change, as well as upon the realities of access to new technologies. We understand more about Lindbergh's flight when we learn how differently African-Americans reacted to Banning and Allen's flight. As a consequence we get better history of technology, better African-American history, and best of all, better American history.

The key to this "new integration," then, is the creation of exhibitions that reflect the interaction among African-Americans and the broader society—a set of transactions that always ran in both directions. These kinds of presentations would explore the clashes, conflicts, compromises, and cultural borrowings that are at the core of the American past. By examining how various peoples throughout history have struggled and negotiated the terms of their lives, exhibits like these can better contextualize the contemporary situations of museum visitors. It is surprising how few effective exhibition models exist. The best of these models is the 1992 exhibition, "Bridges and Boundaries: African-Americans and American Jews," curated by Gretchen Sorin for the Jewish Museum in New York.[12] This exhibition, by examining the shifting sentiments and interaction between African-Americans and American Jews during the twentieth century, demonstrates that in spite of conceptual difficulties and contemporary inter-community tensions, the audience is treated to a richer, more complex, and ultimately more satisfying history.

There are many challenges to crafting effective exhibitions that explore African-American culture. Not the least of which is the many different meanings that Americans can glean from that experience. The museum profession has made great strides in the last decade. If we can continue to take risks, explore difficult questions, and create a more inclusive understanding of the American past, then the museum profession can truly become, in the words of John Cotton Dana, founder and director of the Newark Museum and a celebrated promoter of public access to knowledge, "places of value and service to the communities in which we live."[13]

Notes

1. This essay depends substantially on material previously published in *History News* (50, autumn 1995: 5–9) and in the *Journal of American Culture* (7, 1984: 100–103).

2. Known as the father of black historiography, Woodson created the Association for the Study of African American History and Life.

3. New York *Herald Tribune*, October 11, 1932.

4. The best discussion of this enthusiasm is in Joseph Corn's book *The Winged Gospel: America's Romance with Aviation, 1900–1950* (Oxford University Press, 1983).

5. For an introduction to Garvey, see Cary D. Wintz, ed., *African American Political Thought, 1890–1930* (M. E. Sharpe, 1995). Garvey's Universal Negro Improvement Association and its historical records, are the focus of a special research project at the African American Studies Center at UCLA, information about which is available at www.isop.ucla.edu/mgpp.

6. Any discussion of this era should begin with Alain Locke, *The New Negro: An Interpretation* (Arno, 1978; reprint edition).

7. Pittsburgh *Courier*, October 8, 1932; New York *Age*, October 15, 1932.

8. Pittsburgh *Courier*, November 5, 1932; Baltimore *Afro-American*, October 15, 1932.

9. John W. Ward, "The Meaning of Lindbergh's Flight," *American Quarterly* 10 (1958): 6–13.

10. Cited by Monroe Work, *Negro Yearbook* (New York, 1932), p. 16.

11. *Opportunity*, October 1934, pp. 300–301; *The Crisis*, November 1932.

12. The catalog from that exhibition, edited by Jack Salzman, Adina Beck, and Gretchen Sorin, was published in revised form, but with the same title, by George Brazillier in New York in 1992. See also Gretchen Sorin and Andrew Dolkart, *Touring Historic Harlem: Four Walks in Northern Manhattan* (New York Landmarks Conservancy, 1997).

13. Chalmers Hadley, *John Cotton Dana, A Sketch* (American Library Association, 1943).

A Bibliography of Technology and the African-American Experience

Amy Sue Bix

Over the last three decades or so, the field of the history of technology has expanded tremendously in both breadth and depth. Researchers in the United States and other countries have produced literally thousands of books and articles, spanning a long chronological stretch from ancient civilizations up to the present. The growing sophistication of this work has established the history of technology as a recognized academic field, one firmly grounded in the broad discipline of history as a whole.[1]

Scholars have paid particular attention in recent years to the history of technology in the United States, as evidenced by the appearance of several useful survey texts.[2] In the process, historians of technology have begun building intellectual bridges to other fields, most notably social history, gender history, environmental history, labor history, and economic and business history. Each of those connections has already opened up productive new directions for research and writing, bringing a fresh perspective to the history of American technology. By exploring gender history, for example, investigators have highlighted hidden assumptions about masculinity, femininity, and technological design and use. They have called attention to previously neglected aspects of technology (for example, the brassiere) or underlined the way in which other, seemingly gender-neutral devices have very deeply encoded meanings.[3] Such studies may help serve as inspiration for emerging work on race and the history of technology, in which scholars can explore with a critical eye the political, cultural, and economic conditions within which machines, tools, and methods are created, adopted, adapted, used, or misused.[4]

While much exciting work is already underway, other avenues of research remain relatively unexplored. Recent years have seen impressive developments in the field of African-American history. Informed by new sets of questions, historians have begun to re-evaluate old interpretations of African-American life and the meaning of race, from the period of slavery up through the civil rights era.[5]

To date, the link between the history of technology and African-American history has remained for the most part undiscovered territory. Too often, historians of technology have remained content to focus on mainstream white middle-class society, papering over crucial issues of race, class, and cultural differences. Too often, historians of African-American life have imagined technology as irrelevant, or as a subject limited to stories of black inventors. If such perspectives could be enlarged, if the best African-American history could be integrated with good, contemporary history of technology, both fields would be multiply enriched.

Existing literature provides tantalizing hints of such promise. This essay is designed to survey the books, articles, and other materials already published which treat, in some manner, the subject of technology and the African-American experience. The very process of doing so serves to highlight areas deserving further attention and important topics as yet mostly untouched. But to begin with, it is essential to set out the boundaries of my essay, the scope of its bibliographical conception. Recent years have brought some excellent scholarship on science and the African-American experience. Historians have started to write biographies of black scientists, to examine various constructions of the idea of race, and to reconsider the nature of science itself.[6] While the subjects of science and technology converge and overlap on many occasions, each field ultimately deserves to be considered on its own terms. Anyone interested in technology and the African-American experience can derive valuable information and insights by examining related literature in the history of science. Nevertheless, to keep this essay focused and its length manageable, I have in most instances chosen to leave science-related topics for another opportunity. Similarly, though I treat with what seem to me the most relevant materials in medical history, the large and important body of research dealing directly with the history of health and medicine in the African-American context largely falls outside the immediate scope of this essay.[7]

While this review of the literature thus deliberately excludes certain categories of publications, I have at the same time chosen to include other types of material that relate to a specific definition of technology. For instance, when considering how technology relates to the lives of ordinary men and women, especially African-Americans, it is vital to move beyond the engineer's vision of machines, bridges, and computers. These topics have their place in the bibliography below, but it will serve our purposes much better to consider technology as covering a broad sweep of devices, tools, skills, usage, and experience. Such a conception brings in small-scale craft and domestic technologies, such as blacksmithing, basket weaving, and quilting. It restores the importance of agricultural techniques and farming technologies in our history, factors too easily overlooked at the close of the twentieth century.[8]

At the same time, any examination of the African-American community's relationship to computers, medical devices, and other modern technologies tends to shade over into the fields of political science, sociology, and policy studies. The topic of technology and African-American life is also one which carries a great deal of interest outside university courses and academic conferences. Popular magazines frequently run brief pieces on "great black inventors" or "African-American computer users." The bibliography below includes a selection of such articles and popular books, identifying ones that seem more relevant to our immediate examination. As a final note, writers in recent years have begun turning out many juvenile books telling the story of black inventors and scientists; for the most part, the list below omits such works.

For easy reference, I have subdivided the subject of technology and the African-American experience into eleven sections:

General Surveys and Reference Works
Inventors and the History of Invention
Agriculture and Farming Technologies
Crafts and Household Technologies
Manufacturing, Workplace Technics, and Labor History
Technology, Economics, and Urban Life
Naval and Aviation History
Technology, Race, and the Environment
Race, Medical Technology, and Health Issues
Computer Technology and Information-Age Access
Technical Education, Employment, and Professional Issues

I will take up each topic in turn, calling attention to some specific sources, describing their scope, and pointing to areas for further exploration. This treatment by topic is followed by an alphabetized list of the publications referred to, along with others worthy of note.

General Surveys and Reference Works

General surveys on African-American history and culture, while not concentrating exclusively on technology-related subjects, may nonetheless contain much valuable information for students and researchers. The five-volume *Encyclopedia of African-American Culture and History* (1996) presents an especially wide-ranging synthesis; the comprehensiveness of its scope makes it a strong starting point for further reading. Jeffrey Steward's *1001 Things Everyone Should Know about African American History* (1996) offers a much briefer treatment, but contains some good beginning material in its chapter on the history of black invention, science, and medicine. *The African American Mosaic* (1994) may also be of service, especially for those new to the field. Readers may also wish to consult *Africana* (Appiah and Gates 1999); this reference work generally offers less breadth on black American history than on the continent of Africa itself, but it has good introductory pieces on subjects ranging from food, music, and material culture to black inventors' lives. *The Harvard Guide to African-American History* (Higginbotham et al. 2001) compiles extensive bibliographies organized by subject; headings include education, work, thought and expression, medicine and health, and science and technology. To facilitate bibliographical searches, a companion CD-ROM accompanies the Guide. The two-volume reference book *Historical Statistics of Black America* contains an amazing number and range of tables, charts, graphs, and summaries on more than three centuries of African-American life, data on black education, population, and living patterns, and economic statistics.

Inventors and the History of Invention

One of the earliest works cited in this bibliography, Henry Baker's *The Colored Inventor* (1915), is well worth looking at as a historical piece. The 1970s' climate of rising attention to black culture brought a small wave of publications on the history of invention. General Electric sponsored *Black Americans in Science and Engineering* (ed. E. Winslow, 1974), Louis Haber's *Black Pioneers of Science and Invention* (1970), and Aaron Klein's *The Hidden Contributors* (1971). With a more specialized focus, the U.S. Department of Energy published *Black Contributors to Science and Energy Technology* (1979). In 1983, Autumn Stanley argued in "From Africa to America" that black women inventors have been twice overlooked by standard histories because of both their race and gender. Her 1995 book *Mothers and Daughters of Invention* pulled out this vanished history of invention for women of all races. In two articles, one published in 1984 and one in 1991, Edward Jenkins examined the social conditions behind black science and invention.[9]

More recent years have brought some works of special note on the history of African-American invention. Portia James's catalog for an exhibit at the Smithsonian Institution's Anacostia Museum includes an excellent survey of the subject. James's *The Real McCoy* (1989), a revised version of which is included in the present volume, covers a span from African people's early arrival on the American continent through the first third of the twentieth century. James's book contains a wealth of visual material, including original patent drawings and photographs of inventions. In 1993, James Michael Brodie published *Created Equal*. Although this popularly oriented book can prove frustrating to researchers because of the absence of footnotes and other scholarly detail, it does contain a handy summary of African-American inventions and patents. Similarly, Raymond Webster's volume *African American Firsts in Science and Technology* (2000) contains chronologically organized notes on inventors (organized into categories such as agriculture and everyday life, mathematics and engineering, and transportation). Though entries may be short and sometimes difficult to verify historically, this reference work taken as a whole can open the way for further research and discussion.

In addition to such surveys, readers may wish to consult the small body of publications about the lives and efforts of individual inventors. The outstanding work in this area, *Blueprint for Change*, edited by Janet Schneider and Bayla Singer (1995), presents three well-written and well-illustrated essays, one on Lewis Latimer's personal biography, one on the historical context of his invention and one on his technical approach. Rayvon David Fouche's 1997 Ph.D. dissertation, The Dark Side of American Technology: Black Inventors, Their Inventions, and the African-American Community, 1875–1925, is the latest contribution to this field of writing. With an analysis well grounded in the history of technology, Fouche reviews the inventive careers of Latimer, Granville T. Woods, and Shelby J. Davidson, analyzing their motives for invention, their technological experiences, and their significance in terms of broader African-American history. The life of shoe-making pioneer Jan Matzeliger has attracted the interest of Sidney Kaplan (1955) and Dennis Karwatka (1991). M. C. Christopher (1981)has written on Granville T. Woods. In two works whose subjects straddle the boundaries between technology and science, Silvio Bedini (1972) has presented a classic biography of the instrument and clockmaker Benjamin Banneker. Linda McMurry (1981) has written on the agriculturalist George Washington Carver. Much work, however, remains to be accomplished in this area; thorough biographies of major black inventors, their approach derived from the latest scholarship in the history of technology, would fill a real need.

The specialized subject of patent law and practice has attracted some useful work, especially as published in the *Journal of the Patent Office Society*. Articles by John Boyle (1960), Patricia Carter-Ives (1975 and 1980), and Dorothy Cowser Yancy (1984) illustrate, among other things, the obstacles that the patent procedure presented for African-American inventors.

Agriculture and Farming Technologies

Thanks in large part to the journal *Agricultural History*, a number of articles on various aspects of African-American agricultural knowledge and farming techniques have been published.[10] For example, Valerie Grim (1994, 1995) has written about the impact of farm mechanization on black farm families in the rural South and about how the growing power of agribusiness affected African-American farmers. Along similar lines, Bonnie Lynn-Sherow (1996) has studied the historical connections between mechanization, land use, and ownership in Oklahoma. For an alternate interpretation of the effects of mechanization, see the recent work of Donald Holley (2000).[11]

The land-grant colleges established by the Morrill Act of 1890 to serve African-American students and communities played a major role in shaping and spreading agricultural and technical skill. In 1991, *Agricultural History* devoted a special issue to that particular subject, with articles both on the general history of land-grant college development and on specific schools such as Tuskegee, Arkansas AM&N, Florida A&M, and South Carolina State.

Although most do not focus primarily on the history of technology, the multitude of books in print on the history of plantation life, slavery, and accompanying system of agriculture provide a wealth of background material. Practicality prevents including all such works in this bibliography, but two with special relevance to readers here are Pete Daniel's 1985 work on cotton, tobacco, and rice cultures, and David Whitten's 1981 look at the life of a black sugar planter in antebellum Louisiana. The history of rice growing in the South, with its demand for special cultivation skills, gives a special perspective on slave labor and African-American culture; on that subject, see Judith Ann Carney's *Black Rice* (2001), Carney's essay in the present volume, and Julia Smith's *Slavery and Rice Culture in Low Country Georgia* (1992). With particular reference to the history of technology specifically, see Angela Lakwete's Ph.D. dissertation on cotton gins and ginning. A useful work connecting the history of plantation life to questions about race and the built environment is John Michael Vlach's *Back of the Big House* (1993).

Crafts and Household Technologies

In recent years, the history of "women's technology" has begun to take its place among the literature on industrial engineering, machine building, and other centers of masculine culture. With this broadening of views, books generally classified as "art history" or "material culture" or "archaeology" deserve to be considered simultaneously for their value to historians of technology. The place to begin is with the bibliography Theodore Landsmark has compiled and deposited in the Henry Francis du Pont Winterthur Library at Wilmington, Delaware. More particularly, admirers of quilt making have produced several valuable books discussing the design, sewing, and significance of African-American quilts. The volume *African-American Quilt-making in Michigan*

(1997) may be of interest, along with works by Roland Freeman (1996), Floris Cash (1995), and Cuesta Benberry (1992). *Hidden in Plain View*, by Jacqueline Tobin and Raymond Dobard (1999), suggests an interesting if controversial thesis linking quilt design to coded Underground Railroad messages for assisting escaping slaves. The subject of basket weaving has attracted attention from Dale Rosengarten (1986) and Mark Wexler (1993), whose work connects the history of craftsmanship to plantation history, African traditions, and the problems created by modern land development. For a broader sweep, readers may wish to consult John Michael Vlach's 1991 book *By the Work of Their Hands*, which contains essays on eighteenth-century blacks' domestic artifacts, on black builders and houses, and on nineteenth-century craft traditions, and also a number of biographical sketches of artisans. Two older survey works on African-American craft and decorative arts, one by Vlach (1978) and one by Judith Wragg Chase (1971), may prove very useful. Nevertheless, there remains a great need for concentrated scholarly attention here, for works that consciously bridge the gaps between gender history, African-American history, and the history of craft skill. Exciting research, for instance, remains to be done on a rigorous analysis of African-American food traditions and cooking technologies. This bibliography does not attempt to list all the African-American heritage cookbooks, many of which include historical photos and personalized narratives (along with mouth-watering recipes). For an analysis of what two such cookbooks authored by black women reveal about their historical understanding and assumptions, see Rafia Zafar's 1999 article "The Signifying Dish."[12]

Starting from a broad definition of technology, some exciting work has emerged from recent attempts to link the history of body and appearance to questions about the evolving meanings attached to race and gender. In *The Body Project* (1997), Joan Jacobs Brumberg discusses the history of how young women have thought about and attempted to change the appearance of their body and skin. Kathy Peiss's *Hope in a Jar* (1998) goes still further, connecting the history of beauty devices and skills to the history of business, as in the case of cosmetic entrepreneur Madam C. J. Walker. While Peiss integrates issues of race throughout her book, the chapter "Shades of Difference" pays special attention to the development of black cosmetics and consumerism. There are several popular books on African-American hair styling and its history; for one analysis more firmly rooted in historical context, see the 1995 article by Shane White and Graham White. These two authors have also written on the history of African-American clothing styles (1998); for more on that subject, see Helen Foster's 1997 book on the production of textiles in the antebellum South and the evolution of dress as a signifier of multiple roles.

The history of craftsmanship in general has stimulated some worthwhile research. *The Other Slaves* (Newton and Lewis 1978) and *American Artisans* (Rock et al. 1995) discuss the role of artisans and enterprise in the Southern slave economy. Archaeological studies of African-American homes and communities have much to offer a historian of technology; as excellent sources, see Laurie Wilkie (2000) on Louisiana plantation living and Barbara Heath (1999) on slave life in rural Virginia. The volume *"I, Too, Am America"* (Singleton 1999) contains articles discussing specific types of artifacts (pipes, pottery, etc.) as well as methodological and philosophical pieces on how archaeologists can understand African-American life; *The Archaeology of Slavery and Plantation Life* also covers a lot of ground. Taking a more focused approach, the Texas Folklore Society has published an interesting piece on African-American blacksmithing by Richard Allen Burns. Catherine Bishir's 1984 work on black builders in antebellum North Carolina is well complemented by George McDaniel's *Hearth and Home* (1982), which brings race into a discussion of buildings, housing, and material culture. Several historians have produced articles discussing individuals who worked as carpenters, tool makers, marble cutters, and other types of craftsmen; see the work of Greg Koos and Marcia Young (1993), Patricia Brady (1993), and Richard DeAvila (1993). The career of North Carolina furniture maker Thomas Day has been covered by Rodney Barfield (1975) and Patricia Marshall (2001). In their stunningly illustrated book *Reflections in Black*, Deborah Willis and Robin Kelley (2000) have documented the history of African-American photography as an art, business, and technology.

There remains much work to be done in order to explore the relationship between the history of technology and consumerism for African-American populations. Recent years have seen some excellent work at the intersection of what Ruth Schwartz Cowan calls the "consumption junction," yet most histories of consumerism and technology assume an audience of white purchasers and users, deferring the issue of race.[13]

Manufacturing, Workplace Technology, and Labor History

In terms of sheer volume, the subject of technology as related to the African-American working experience has drawn the greatest amount of attention of all the topics covered in this essay. Historians of technology in recent years have been striving to connect their work to labor history, opening up new questions about the nature of industrial enterprise, manufacturing, and other forms of work. The literature here is uneven; some books and articles actively foreground issues of technology and workplace machinery, while others remove such questions to the background. Still, such works can provide a serviceable starting point for anyone wishing to pursue research at the place where the history of technology meets labor history and African-American history.[14]

The most comprehensive point from which to start examining the history of black labor is undoubtedly *Black Workers* (Foner and Lewis 1989), which provides a wealth of primary source material from the course of two centuries, reproducing documents that illustrate the history of African-American work in shipyards, stockyards, train yards, lumbering, sharecropping, automobile factories, and elsewhere. Concentrating more exclusively on early twentieth-century manufacturing, *African-Americans in the Industrial Age* (Trotter and Lewis 1996) contains extensive information drawn from oral histories, newspaper reporting, employment records, and other written sources. As a starting point to a broader analysis of how race relations and racial attitudes played a role in defining working-class identity, see the discussion of white and black longshoremen, steelworkers, and unionization in Bruce Nelson's *Divided We Stand* (2001).

Again, while limitations of space preclude the idea of listing everything, books and articles discussing general history of the pre-Civil War South can supply useful background information on slavery and techniques of work. A couple older treatments still worthy of note here are Robert Perdue's *Black Laborers and Black Professionals in Early America, 1750–1830* (1975) and Robert Starobin's *Industrial Slavery in the Old South* (1970). More recently, Charles Dew (1994) has written on slave-based iron-making and agriculture in Virginia, and Ronald Bailey (1994) has examined the place of black labor in the processes of cotton production, and textile-based industrialization. One book providing a unique perspective on this history of African-American work and invention is the 1971 reprint of Giles B. Jackson and D. Webster Davis's book *The Industrial History of the Negro Race of the United States*, originally written in 1908 in connection with the Jamestown Tercentennial Exposition.

Dennis Dickerson (1986) has traced the history of black steelworkers in Pennsylvania since 1875. Henry McKiven's work on the racial dimensions of Birmingham's iron and steel industry (1995) is similarly useful, as is W. David Lewis's work on the Sloss furnaces (1994). Ronald Lewis (1987), Joe William Trotter (1990), and Daniel Letwin (1998) have all written on the history of race, class, community, and unionism in coal mining. Recently, the subject of race, labor relations, and the division of work in the meat packing industry has attracted attention from Roger Horowitz (1997) and Rick Halpern (1996, 1997). Eric Arnesen (2001) addresses the history of railroads' reliance on black labor, describing how African-Americans working as locomotive firemen, brakemen, and porters attempted to organize and fight discrimination. Earlier work by Arnesen (1991) discusses black waterfront workers and unionization in New Orleans from the Reconstruction Era through the early twentieth century. Lisa McGirr (1995) has discussed racial attitudes in longshoremen's work, a subject also treated in *Waterfront Workers* (Winslow 1998). August Meier (1979) and Kevin Boyle (1995,

1997) have explored the history of African-Americans' roles in automobile manufacturing and the United Auto Workers union. Samuel James (1985) has considered how robotization in the auto industry might affect black laborers. In *Race on the Line*, Venus Green (2001) has compiled a detailed history of job segregation based on race and gender in the telephone industry, analyzing how introduction of new switchboard technologies affected workers' skills, control, and employment. E. Valerie Smith (1993) has examined the part blacks played in building the Alaska Highway; Michael Conniff (19485) has fulfilled a similar goal in regard to the Panama Canal.

The entire subject of African-Americans' long experience not only as employees but also as businessmen has interested a number of historians. Whittington Bernard Johnson (1993) has looked at the development of black labor and business enterprise in the period from 1750 to 1830. Gavin Wright (1986) and Jay Mandle (1992) have looked at issues of race and economic experience in the decades that followed the Civil War. Robert Margo (1990) has looked at the economic implications of a racialized school system in the South. Alexa Henderson (1990) has investigated the relationship between business and Atlanta's black middle class. A good comprehensive work is Juliet Walker's *History of Black Business in America* (1998). The *Encyclopedia of African-American Business History* (Walker 1999) has entries on antebellum trades and businesses, nineteenth-century black-owned factories and manufacturing ventures, and twentieth-century telecommunications companies and inventors. Martin and Jacqueline Hunt (2000) offer an overview of black entrepreneurship and business development that includes more than a dozen case studies of large black-owned corporations. A'Lelia Bundles (2001) gives a detailed account of the fascinating life of her great-great-grandmother Madam C. J. Walker, who became well known for building up a business empire and sales network which distributed hair-care products for African-American women. Jonathan Greenberg (1990) has taken up the case of an African-American petroleum business. Historians of technology may be able to extract much useful information from such publications in economic and business history.

Technology, Economics, and Urban Life

The subject of African-Americans' history as workers leads naturally to the topic of technology, economics, and urbanization. Books tracing the racial dimensions of the development of American cities include *Black Communities and Urban Development in America, 1720–1990* (1991), *The Metropolis in Black and White* (1992), *The New African-American Urban History* (1996), and *Historical Roots of the Urban Crisis* (2000).[15] Nicholas Lemann's *The Promised Land* (1991) outlines the history of the "great black migration" in relation to individuals, families, and communities; readers might also want to look at Craig Heinicke's 1994 article on population movement and urban labor skills. Steven Hoffman (1993) has written on race and the building of Richmond, Virginia, Kenneth Kusmer (1978) on Cleveland, Henry Louis Taylor Jr. (1986) on Cincinnati, and Michael Fitzgerald (1993) on Mobile. Robert Johnson (1984) has looked specifically at the topic of "Science, Technology, and Black Community Development." Melvin Mitchell (2001) has combined a look the history of city development with a critical commentary on black architectural professionals and a call for a New Black Urbanism.

The process of designing and constructing municipal services and systems carried important implications for race, class, and community life. Mark Rose (1995) has written about such issues in the history of utility development in Denver and Kansas City, and Ron Nixon (1995) has noted the case of a lingering racial divide in access to water supplies. There is surprisingly little scholarly research on how the development of automobiles affected African-American life, but Howard Preston has done significant work outlining racial aspects of Atlanta's automobile-age development from 1900 to 1935. In his classic essay "Do Artifacts Have Politics?" Langdon Winner (1986) drew out the way Robert Moses built racial and class lines into his design for Long Island parkways; those interested in the fascinating career of Moses may also wish to consult works by Robert Caro (1974) and Joel Schwartz (1993). More in the realm of policy than in the realm of pure history, *Just Transportation* (Bullard and Johnson 1997) discusses the shape of modern transport systems. Nancy Grant's *TVA and Black Americans* (1990) connects the history of New Deal government planning and regional economic development to ideas of society and race.

The subject of race and urban life leads to the history of the twentieth-century amusement business, particularly the rise of movie technology. Douglas Gomery (1992), David Nasaw (1993), and Gregory Waller (1995) have written on the modernization of the urban environment and the history of racial segregation in the theater business. Examining the link between history of technology and African-American popular culture, Tricia Rose (1994) has discussed samplers, drum machines, and other technologies as factors in the emergence of rap music.[16]

Naval and Aviation History

Again, in terms of sheer volume, the topic of African-Americans' place in military history has drawn substantial attention, both academic and popular.[17] Subtopics most likely to prove relevant to historians of technology include naval history and aviation history. Michael Cohn and Michael Platzer (1978), Martha Putney (1987), James Barker Farr (1989), and W. Jeffrey Bolster (1990, 1997) have all discussed the history of African-Americans as sailors, merchant seamen, and whalers. The Smithsonian Institution Press has performed a leading role in publishing books by and about blacks in aviation, including the autobiographical works of William Powell (1994, originally issued in 1934), of experimental-airplane builder Neil Loving (1994), and of Janet Harmon Bragg (1996). Doris Rich has performed a valuable service in her book on Bessie Coleman, *Queen Bess* (1993), marrying the history of race to the history of gender in aviation. Among the numerous works on the history of the Tuskegee Airmen, Charles Walter Dryden's memoir (1997) and Stanley Sandler's *Segregated Skies* (1992) stand out. For anyone with a particularly strong interest in this area, Betty Kaplan Gubert has compiled a bibliography, *Invisible Wings* (1994), detailing numerous works in the history of black aviation. While at times more anecdotal than academic, *They Had a Dream* by J. Alfred Phelps (1994) recounts the difficult story of blacks in the United States astronaut corps. Jill Snider's 1995 University of North Carolina dissertation, Flying to Freedom, provides a cultural history approach to the subject.

Technology, Race, and the Environment

Since the 1980s there has been a surge of discussion, theorizing, and writing on the subjects of "environmental justice" and "environmental racism." Topics of concern include the pollution of poor and minority communities in relationship to the siting of factories, toxic-waste dumps, and other environmental hazards, with potentially devastating implications for health and the quality of life. Luke Cole and Sheila Foster (2000) have documented the rise of the environmental justice movement, detailing key episodes of community protest and grassroots activism across the United States. As a general lead-in to the concept of environmental justice itself, readers may do best to look at the publications of Robert Bullard (1992), Alice Brown (1993), Robert Collin (1993), and Robert Knox (1993). For more, see the National Academy of Science's report *Toward Environmental Justice* (1999); see also *Environmental Injustices, Political Struggles* (Camacho 1998); *Unequal Protection* (Bullard 1994), *Race and the Incidence of Environmental Hazards* (Bryant and Mohai 1992), *Confronting Environmental Racism* (Bullard 1993), and *Environmental Justice: Hearings before the Subcommittee on the Judiciary* (1994). Dealing still more immediately with the race and class dimensions of hazardous waste are works by Francis Adeola (1994), Robert Bullard (1990), and *Toxic Wastes and Race in the United States* (1987). Andrew Hurley (1995) has taken a more historically based look at patterns of industrial pollution in

Gary, Indiana in the years after World War II. Ballus Walker (1991) has emphasized the implications for public health.

Race, Medical Technology, and Health Issues

The twentieth century has seen a growing overlap between the field of medicine and the development of technology. While the broad subject of race and medicine falls mostly outside the immediate purview of this essay, the two-volume work *An American Health Dilemma* by W. Michael Byrd and Linda Clayton (2000, 2001) provides an excellent introduction to the history of black healers, medical racism, and medical inequality, and African-American health-care issues.[18] Other good basic sources include the works of Edward Beardsley (1992) and Nancy Krieger and Mary Bassett (1993). Beardsley has also written (1987) on the history of health care for the Southern black population, a matter that leads inevitably to the history of the Tuskegee syphilis experiment. James Jones's 1993 book *Bad Blood* remains a strong source to consult on that miserable chapter of history. *Tuskegee's Truths* (Reverby 2000) contains excerpts from primary documents and interviews with involved parties, along with a large number of excellent essays revisiting the medical history, the racial legacy, and the legal and bioethical morals that are to be derived. While such discussions may not involve "technology" in its narrowest sense as machines and tools, they do provide a strong lead-in for considering questions about the historic exclusion of blacks, other minorities, and the poor from an era's most sophisticated medical understanding. The issue of expertise as contested terrain is an important one, and the history of medical technology is one of empowering its possessors and unfortunately, occasionally victimizing the subjects.

On other specific topics, Melbourne Tapper (1999) and Keith Wailoo (1997, 2001) provide thought-provoking material on the racialized construction of diseases such as sickle-cell anemia. In *One Blood* (1996), Spencie Love offers a fascinating account of how perceptions of racial discrimination in medical care became written into popular folklore.

Readers who wish to move from history toward the realm of policy studies enjoy an overabundance of sources. *Health Policies and Black Americans* (1989) provides one starting point, as do the books of Eric Bailey (1991), Wornie Reed (1992), and Susan Smith (1995). From the early 1990s on, various departments of the federal government and Congressional committees produced numerous reports on various questions about minority health. However, readers must be careful when selecting such sources, since policy material can quickly become dated.

In recent years, observers have begun paying more attention to different racial patterns in the occurrence and treatment of specific diseases. For a few examples of such studies, see the articles by Sandra Blakeslee (1989), George Friedman-Jimenez (1989), and David Brown (1992). The *Journal of the American Medical Association*, the *New England Journal of Medicine,* and other medical publications continue to carry reports about racialized differentials in heart-disease treatment, cancer therapy, and overall patient care.

The subject of race and medical care has also become entangled with questions about ethics. Questions about the African-Americans' stake in the moral dimensions of medical research appear in the works of Gina Kolata (1991) and Isabel Wilkerson (1991) and in *Beyond Consent* (Kahn et al. 1998). Annette Dula (1991) discusses "An African-American Perspective on Bioethics." A 1996 volume titled *The Human Genome Project and the Future of Health Care* contains an essay on how the mapping of the human genome might affect health service for minority populations. Issues of race have also entered writings about practices and technologies of birth control, as can be seen in the works of Jessie Rodriquez (1989) and Dorothy Roberts (1997). On how to incorporate racial dimensions into the teaching of medical history and medical policy, see Vanessa Gamble (1999).

Computer Technology and Information-Age Access

With the ever-accelerating power of computer technology to transform work, recreation, and behavior, observers have raised powerful questions about the implications of differential access to information-age life.[19] The 1990s brought a series of studies, articles, and surveys exploring the issue of whether or not there is a serious class and race-related gap in computer use. Discussion of the "digital divide" has been controversial, with some observers suggesting that there remains a stubborn and sizable gulf between those Americans who regularly use new information technologies and those who do not; others maintain that any such separation is rapidly and naturally vanishing. Authors who have taken different angles on that debate include Michel Marriott (1995), Deidra Ann Parrish (1997), and Salim Muwakkil (1998). A 1998 article by Donna Hoffman and Thomas Novak highlighted evidence of racial differentials in computer use. Articles by Joachim Krueger (1998) and Frederick McKissack (1998) follow up on the implications of the *Science* study. Other works on the subject include *Cyberghetto or Cybertopia?* (Bosah 1998) and *Disconnected* (Wresch 1996). The latest literature includes the work of David Bolt and Ray Crawford (2000), which accompanies a PBS series documenting how children experience a difference in home and school-based technology resources and describing organizations which attempt to correct this problem of differential availability. *The Digital Divide* (2001) offers data along with an interpretation suggesting the closing of the gap. A policy-oriented analysis discussing education and grassroots initiatives (centered around access for poor communities, rather than race per se) can be found in *High Technology and Low-Income Communities* (Schon et al. 1998).

As Evelyn Nakano Glenn and Charles Tolbert II (1987) have noted, access to technical education has serious ramifications for employment, with computers intensifying divisions among workers by race, education, and skill. For those interested in historical analogies, Richard Sclove and Jeffrey Scheuer (1994) offer a popular analysis comparing the information superhighway to the construction of interstate highways in earlier decades. For those interested in policy activity, Bennett Harrison (1997) has written about cooperative efforts between businesses and community groups to offer technology-related job training to poor urban populations.

In recent years, *Black Enterprise* has periodically published articles and special issues devoted to the role of technology in African-Americans' business, education, and everyday life. Tariq Muhammad (1998) and Roger Crockett (1998) comment on ways in which the Internet may open and may close opportunities for black entrepreneurs and technical experts. Such cheerleading articles in a business-oriented magazine sometimes lack a critical perspective, yet nonetheless may prove useful for certain functions. Historians may do well to balance such compensatory promotional pieces against other sources, integrating them with a more scholarly approach.

Those interested in the recent history of technology as a vehicle for expressing African-American culture in the information age may find some thought-provoking if incomplete lines of research. Ron Eglash (1995), in his cultural analysis of the race-related dimensions of cybernetic theory, views music styles and other forms of vernacular expression as a clash of analog and digital representations.

Technical Education, Employment, and Professional Issues

The subject of technology and the African-American experience is connected to an entire range of history involving scientific and engineering education, employment, and professionalization. James Anderson has written extensively on the subject. For a broad coverage of the topic, see his 1988 book *The Education of Blacks in the South, 1865–1935*.[20] Nina Lerman (1997) has explored race and gender issues in the establishment of technical education in mid-nineteenth-century Philadelphia. Clyde Woodrow Hall (1973) has written on the history of vocational and technical education for African-Americans. David Wharton's 1992 book *A Struggle Worthy of Note* has chapters detailing the history of black colleges and engineering training. (For a contrasting view, see Donald Spivey's 1978 book *Schooling for the New Slavery*.)

A work of special interest on the subject of black engineering education has grown out of the Blacks at MIT History Project. Clarence Williams's *Technology and the Dream* (2001) compiles more than 75 oral history interviews in which black students, faculty, and staff discuss their educational experiences, choices, successes and frustrations working at the Massachusetts Institute of Technology over the period of half a century.[21]

In recent years, the National Science Foundation, the Office of Technology Assessment, and many other organizations have periodically published reports pointing out the latest rise or fall in minority enrollments in science and engineering. As examples, readers may wish to look at *Higher Education for Science and Engineering* (1989) and at the works of Susan Hill (1992) and Constance Holden (1995). Public and private groups will undoubtedly continue to survey the situation and issue bulletins over upcoming years. Authors of articles discussing strategies for improving minority science and engineering education include Charles Farrell (1988), Ward Worthy (1990), Larry De Van Williams (1990), Calvin Sims (1992), C. C. Campbell-Rock (1992), Norman Fortenberry (1994), Jerike Grandy (1997), and Mary Anderson-Rowland et al. (1999). Juan Lucena (2000) has discussed the philosophy behind identification and concentrated education efforts for minorities.

The 1990s brought a series of reports detailing and commenting on the latest trends in the employment and professionalization of African-American engineers and scientists. The American Association for the Advancement of Science published *Investing in Human Potential* (Matyas and Malcolm 1991). In 1992 the NSF presented a report by Patricia White titled *Women and Minorities in Science and Engineering: An Update*, and the magazine *Science* summarized emerging conclusions in an article titled "Minorities in Science: The Pipeline Problem." Other useful sources relating to the "pipeline" issue include articles by Kay Whitmore (1998), Sue Kenmitzer (1988), Leonard Waks (1991), Shirley Malcolm (1993), and Richard Stone (1993). Anyone who would like leads to still more sources of information may consult the 1989 Library of Congress publication by Vivian Sammons and Denis Dempsey titled *Blacks in Science and Related Disciplines*.

Summing It All Up

The above account ultimately serves to underline two related points. On the one hand, recent years have brought the publication of many significant books, articles, and reports in the areas of history and policy. Anyone willing to explore the stacks in a library can find substantial material (of uneven quality) treating "technology and the African-American experience." And there is a surprising amount of material on the World Wide Web. The Louisiana State University Library site, for instance, has extensive information on black inventors and scientists. On the other hand, the gaps in this literature remain at least as visible as the conclusions. Scholars face much work ahead in any attempt to bring the history of technology together with African-American history in a way that does full justice to both, but the effort promises to yield rich new intellectual dividends.

Alphabetical Listing of Books, Articles, and Other Materials[22]

Adeola, Francis O. 1994. "Environmental Hazards, Health, and Racial Inequity in Hazardous Waste Distribution." *Environment and Behavior* 26, January: 99–126.

Agricultural History. 1991. Special issue on the history of black land-grant agricultural and engineering colleges (volume 65, spring).

Anderson, James D. 1988. *The Education of Blacks in the South, 1869–1935*. University of North Carolina Press.

Anderson-Rowland, Mary R., Stephanie L. Blaisdell, Shawna L. Fletcher, Peggy A. Fussell, Mary Ann McCartney, Maria A. Reyes, and Mary Aleta White. 1999. "A Collaborative Effort to Recruit and Retain Under-represented Engineering Students." *Journal of Women and Minorities in Science and Engineering* 5: 323–350.

"A Place at the Table: A Sierra Roundtable on Race, Justice, and the Environment." 1993. *Sierra* 78, May-June: 50ff.

Appiah, Kwame Anthony, and Henry Louis Gates, eds. 1999. *Africana: The Encyclopedia of the African and African-American Experience*. Basic Civitas Books.

Arnesen, Eric. 1991. *Waterfront Workers of New Orleans: Race, Class, and Politics, 1863–1923*. Oxford University Press. Paperback reprint: University of Illinois Press, 1994.

Arnesen, Eric. 2001. *Brotherhoods of Color: Black Railroad Workers and the Struggle for Equality*. Harvard University Press.

Bailey, Eric J. 1991. *Urban African American Health Care*. University Press of America.

Bailey, Ronald. 1994. "The Other Side of Slavery: Black Labor, Cotton, and the Textile Industrialization of Great Britain and the United States." *Agricultural History* 68, spring: 35–50.

Baker, Henry E. 1915. *The Colored Inventor: A Record of Fifty Years*. Reprint: Arno, 1969.

Barfield, Rodney. 1975. *Thomas Day, Cabinetmaker*. North Carolina Museum of History.

Beardsley, Edward H. 1987. *A History of Neglect: Health Care for Blacks and Mill Workers in the Twentieth-Century South*. University of Tennessee Press.

Beardsley, Edward H. 1992. "Race as a Factor in Health." In *Women, Health, and Medicine in America*, ed. R. Apple. Rutgers University Press.

Bedini, Silvio A. 1972. *The Life of Benjamin Banneker*. Scribner.

Benberry, Cuesta. 1992. *Always There: The African-American Presence in American Quilts*. Kentucky Quilt Project.

Berlin, Ira, and Phillip Morgan, eds. 1990. *The Slaves' Economy: Independent Production by Slaves in the Americas*. Frank Cass.

Betancur, John J., and Douglas C. Gills. 1993. "Race and Class in Economic Development." In *Theories of Local Economic Development*, ed. R. Bingman and R. Mier. Sage.

"Beyond White Environmentalism: Minorities and the Environment." 1990. *Environmental Action*, January-February: 19–30.

Billingsley, Andrew. 1988. "The Impact of Technology on Afro-American Families." *Family Relations* 37, October: 420–425.

Bishir, Catherine. 1984. "Black Builders in Antebellum North Carolina." *North Carolina Historical Review* 61, October: 422–461.

"Black Adults Among Increased Internet Users." 1998. *Jet* 94, September 14: 36.

"Black Colleges Holding Their Own in Preparing Students for Doctorates in the Sciences." 1997. *Journal of Blacks in Higher Education*, summer: 58–59.

Black Contributors to Science and Energy Technology. 1979. Office of Public Affairs, U.S. Department of Energy.

Black Enterprise. 1996. Special issue on African-American business and technology (volume 26, March).

"Blacks in Aviation History." 1994. *Ebony* 49, February: 118.

Blakeslee, Sandra. 1989. "Studies Find Unequal Access to Kidney Transplants." *New York Times*, January 24.

Bolster, W. Jeffrey. 1990. "'To Feel Like a Man': Black Seamen in the Northern States, 1800–1860." *Journal of American History* 76, March: 1173–1198.

Bolster, W. Jeffrey. 1997. *Black Jacks: African American Seamen in the Age of Sail*. Harvard University Press.

Bolt, David B., and Ray A. K. Crawford. 2000. *Digital Divide: Computers and Our Children's Future*. TV Books.

Boyle, John. 1960. "Patents and Civil Rights in 1857–8." *Journal of the Patent Office Society* 42: 789–794.

Boyle, Kevin. 1995. "'There Are No Union Sorrows That the Union Can't Heal': The Struggle For Racial Equality in the United Automobile Workers, 1940–1960." *Labor History* 36, winter: 5–23.

Boyle, Kevin. 1997. "The Kiss: Racial and Gender Conflict in a 1950s Automobile Factory." *Journal of American History* 84, September: 496–523.

Brady, Patricia. 1993. "Florville Foy, F.M.C.: Master Marble Cutter and Tomb Builder." *Southern Quarterly* 31, winter: 8–20.

Bragg, Janet Harmon. 1996. *Soaring Above Setbacks: The Autobiography of Janet Harmon Bragg, African American Aviator*. Smithsonian Institution Press.

Brodie, James Michael. 1993. *Created Equal: The Lives and Ideas of Black American Innovators*. William Morrow.

Brown, Alice L. 1993. "Environmental Justice: New Civil Rights Frontier." *Trial* 29, July: 48–53.

Brown, David. 1992. "Racial Disparity Found in Bypass Surgery Rate." *Washington Post*, March 18.

Brumberg, Joan Jacobs. 1997. *The Body Project: An Intimate History of American Girls.* Random House.

Bryant, Bunyan, ed. 1995. *Environmental Justice: Issues, Policies, and Solutions.* Island.

Bryant, Bunyan, and Paul Mohai, eds. 1992. *Race and the Incidence of Environmental Hazards: A Time for Discourse.* Westview.

Bullard, Robert D., ed. 1993. *Confronting Environmental Racism: Voices from the Grassroots.* South End.

Bullard, Robert D. 1994. *Dumping in Dixie: Race, Class, and Environmental Quality.* Westview.

Bullard, Robert D., ed. 1994. *Unequal Protection: Environmental Justice and Communities of Color.* Sierra Club Books.

Bullard, Robert D., and Glenn S. Johnson, eds. 1997. *Just Transportation: Dismantling Race and Class Barriers to Mobility.* New Society.

Bullard, Robert D., and B. H. Wright. 1992. "The Quest for Environmental Equity: Mobilizing the African-American Community for Social Change." In *American Environmentalism*, ed. R. Dunlap and A. Mertig. Taylor and Francis.

Bundles, A'Lelia. 2001. *On Her Own Ground: The Life and Times of Madam C. J. Walker.* Scribner.

Burns, Richard Allen. 1996. "African-American Blacksmithing in East Texas." In *Juneteenth Texas*, ed. F. Abernethy. University of North Texas Press.

Byrd, W. Michael, and Linda A. Clayton. 2000. *An American Health Dilemma*, Volume 1: *A Medical History of African-Americans and the Problem of Race: Beginnings to 1900.* Routledge.

Byrd, W. Michael, and Linda A. Clayton. 2001. *An American Health Dilemma*, Volume 2: *Race, Medicine, Health Care in the United States.* Routledge.

Camacho, David E., ed. 1998. *Environmental Injustices, Political Struggles: Race, Class, and the Environment.* Duke University Press.

Campbell-Rock, C. C. 1992. "African-American Engineers: Turning Dreams into Reality." *Black Collegian* 22, January: 120–123.

Carey, Phillip. 1977. "Engineering Education and the Black Community: A Case for Concern." *Journal of Negro Education* 46: 39–45.

Carney, Judith Ann. 2001. *Black Rice: The African Origins of Rice Cultivation in the Americas.* Harvard University Press.

Caro, Robert A. 1974. *The Power Broker: Robert Moses and the Fall of New York.* Knopf.

Carter-Ives, Patricia. 1975. "Giles B. Jackson, Director-General of the Negro Development and Exposition Company of the U.S. for the Jamestown Tercentennial Exposition of 1907." *Negro History Bulletin* 30, December: 671–682.

Carter-Ives, Patricia. 1980. "Patent and Trademark Innovations of Black Americans and Women." *Journal of the Patent Office Society* 62, February: 108–126.

Carter-Ives, Patricia. 1988. *Creativity and Invention: The Genius of Afro-Americans and Women in the United States and Their Patents*. Research Unlimited.

Carver, Bernard A. 1994. "Defining the Context of Early Computer Learning for African-American Males in Urban Elementary Schools." *Journal of Negro Education* 63, autumn: 532–545.

Cash, Floris Barnett. 1995. "Kinship and Quilting: An Examination of an African-American Tradition." *Journal of Negro History* 80, winter: 30–41.

Cecelski, David S. 2001. *The Waterman's Song: Slavery and Freedom in Maritime North Carolina*. University of North Carolina Press.

Chaplin, Joyce E. 1996. *An Anxious Pursuit: Agricultural Innovation and Modernity in the Lower South, 1730–1815*. University of North Carolina Press.

Chase, Judith Wragg. 1971. *Afro-American Art and Craft*. Van Nostrand Reinhold.

Christopher, M. C. 1981. "Granville T. Woods: The Plight of a Black Inventor." *Journal of Black Studies* 11, March: 269–276.

Clines, Francis X. 2001. "Wariness Leads to Motivation in Baltimore Free-Computer Experiment." *New York Times*, May 24.

Cohn, Michael, and Michael K. H. Platzer. 1978. *Black Men of the Sea*. Dodd, Mead.

Cole, Beverly P. 1985. "Educating Blacks for Tomorrow's Technology." *The Crisis* 92, April: 32–33.

Cole, Luke W., and Sheila R. Foster. 2000. *From the Ground Up: Environmental Racism and the Rise of the Environmental Justice Movement*. New York University Press.

Collin, Robert W. 1993. "Environmental Equity and Need for Government Intervention: Two Proposals." *Environment* 35, November: 41–43.

Committee on Environmental Justice. 1999. *Toward Environmental Justice: Research, Education, and Health Policy Needs*. National Academy Press.

Committee on Increasing Minority Participation in the Health Professions, Institute of Medicine. 1994. *Balancing the Scales of Opportunity: Ensuring Racial and Ethnic Diversity in the Health Professions*. National Academy Press.

Compaine, Benjamin M., ed. 2001. *The Digital Divide*. MIT Press.

Conniff, Michael L. 1985. *Black Labor on a White Canal: Panama, 1904–1981*. University of Pittsburgh Press.

Craig, Lee A. 1992. "'Raising Among Themselves': Black Educational Advancement and the Morrill Act of 1890." *Agriculture and Human Values* 9, winter: 31–37.

Crockett, Roger O. 1998. "Invisible and Loving It: Black Entrepreneurs Find That the Internet's Anonymity Removes Racial Obstacles." *Business Week*, October 5: 124ff.

Cunningham, Patricia A., and Susan Voso Lab, eds. 1993. *Dress in American Culture*. Bowling Green State University Popular Press.

Cutcliffe, S. H., Goldman, S. L., Medina, M., and Sanmartin, Jose, eds. 1992. *New Worlds, New Technologies, New Issues*. Lehigh University Press.

Daniel, Pete. 1985. *Breaking the Land: The Transformation of Cotton, Tobacco, and Rice Cultures since 1880*. University of Illinois Press.

DeAvila, Richard T. 1993. "Caesar Chelor and the World He Lived In." *Chronicle of the Early American Industries Association* 46, December: 91–97.

De Van Williams, Larry. 1990. "Educating Minority Children in an Environment That Makes Engineering Education an Attainable Goal." *IEEE Communications Magazine* 28, December: 58–60.

Dew, Charles B. 1987. *Ironmaker to the Confederacy: Joseph R. Anderson and the Tredegar Iron Works*. Broadfoot.

Dew, Charles B. 1994. *Bond of Iron: Master and Slave at Buffalo Forge*. Norton.

Dickerson, Dennis C. 1986. *Out of the Crucible: Black Steelworkers in Western Pennsylvania, 1875–1980*. State University of New York Press.

Downing, L. K. 1935. "The Negro in the Professions of Engineering and Architecture." *Journal of Negro Education* 4, January: 60–70.

Dryden, Charles Walter. 1997. *A-Train: Memoirs of a Tuskegee Airman*. University of Alabama Press.

Du Bois, W. E. B., ed. 1912. *The Negro American Artisan*. Atlanta University. Reprint: Arno, 1968.

Dula, Annette. 1991. "Towards an African-American Perspective on Bioethics." *From the Center* 10, summer: 1–2.

Ebo, Bosah, ed. 1998. *Cyberghetto or Cybertopia? Race, Class, and Gender on the Internet*. Praeger.

Eglash, Ron. 1995. "African Influences in Cybernetics." In *The Cyborg Handbook*, ed. C. Gray.

Eglash, Ron. 1997. "The African Heritage of Benjamin Banneker." *Social Studies of Science* 27, April: 307–316.

"Engineering Programs at HBCUs Working Just as Hard for Students." 1988. *Black Issues in Higher Education*, October 27.

Environmental Justice: Hearings before the Subcommittee on Civil and Constitutional Rights of the Committee on the Judiciary, House of Representatives, 103rd Congress, 1st Session, Mar. 3–4, 1993. 1994. U.S. Government Printing Office.

Farr, James Barker. 1989. *Black Odyssey: The Seafaring Traditions of Afro-Americans*. P. Lang.

Farrell, Charles S. 1988. "Successful Engineering Schools Recruit Students as Early as Junior High School." *Black Issues in Higher Education*, October 27: 23.

Ferguson, Leland. 1992. *Uncommon Ground: Archaeology and Early African America, 1650–1800*. Smithsonian Institution Press.

Fett, Sharla M. 2002. *Working Cures: Healing, Health, and Power on Southern Slave Plantations*. University of North Carolina Press.

Fitzgerald, Michael W. 1993. "Railroad Subsidies and Black Aspirations: The Politics of Economic Development in Reconstructing Mobile, 1865–1879." *American Heritage* 44, September: 240–256.

Flack, Harley, and Edmund D. Pellegrino. 1989. "New Data Suggest African-Americans Have Own Perspective on Biomedical Ethics." *Kennedy Institute of Ethics Newsletter* 3, April: 1–2.

Flynn, Charles L. 1992. *White Land, Black Labor: Caste and Class in Late Nineteenth-Century Georgia*. Louisiana State University Press.

Foner, Philip S., and Ronald L. Lewis, eds. 1989. *Black Workers: A Documentary History from Colonial Times to the Present*. Temple University Press.

Fortenberry, Norman L. 1994. "Engineering, Education, and Minorities: Where Now." *Journal of Women and Minorities in Science and Engineering* 1: 89–98.

Foster, Helen Bradley. 1997. *"New Raiments of Self": African American Clothing in the Antebellum South*. Berg.

Fouche, Rayvon. 1997. The Dark Side of American Technology: Black Inventors, Their Inventions, and the African-American Community, 1875–1925. Ph.D. dissertation, Cornell University.

Fouche, Rayvon. 1997. "The Exploitation of an African-American Inventor on the Fringe: Granville T. Woods and the Process of Invention." *Western Journal of Black Studies* 21, fall: 190–198.

Freeman, Roland L. 1996. *A Communion of the Spirits: African-American Quilters, Preservers, and Their Stories*. Rutledge Hill.

Friedman-Jimenez, George. 1989. "Occupational Disease Among Minority Workers: A Common and Preventable Public Health Problem." *AAOHN Journal* 37, February: 64–70.

Galster, George C., and Edward W. Hill, eds. 1992. *The Metropolis in Black and White: Place, Power, and Polarization*. Center for Urban Policy Research.

Gamble, Vanessa Northington. 1999. "Teaching about Race and Racism in Medical History." *Radical History Review* 74: 140–161.

Garrity-Blake, Barbara J. 1994. *The Fish Factory: Work and Meaning for Black and White Fishermen of the American Menhaden Fishery*. University of Tennessee Press.

GEM Symposium. 1986. *America Educates for Tomorrow*. National Consortium for Graduate Degrees for Minorities in Engineering.

Gilbert, Charlene, and Eli Quinn. 2000. *Homecoming: The Story of African-American Farmers*. Beacon.

Glenn, Evelyn Nakano, and Charles M. Tolbert II. 1987. "Technology and Emerging Patterns of Stratification for Women of Color: Race and Gender Segregation in Computer Occupations." In *Women, Work, and Technology*, ed. B. Wright. University of Michigan Press.

Goings, Kenneth W., and Raymond A. Mohl, eds. 1996. *The New African-American Urban History*. Sage.

Goldin, Claudia Dale. 1976. *Urban Slavery in the American South, 1820–1860: A Quantitative History*. University of Chicago Press.

Gomery, Douglas. 1992. *Shared Pleasures: A History of Movie Presentation in the United States*. University of Wisconsin Press.

Grandy, Jerike. 1997. "Gender and Ethnic Differences in the Experiences, Achievements, and Expectations of Science and Engineering Majors." *Journal of Women and Minorities in Science and Engineering* 3: 119–144.

Grant, Nancy L. 1990. *TVA and Black Americans: Planning for the Status Quo*. Temple University Press.

Graves, Earl G. 1998. "Staking a Claim on the Future." *Black Enterprise* 28, March: 11.

Green, Venus. 1995. "Race and Technology: African-American Women in the Bell System, 1945–1980." *Technology and Culture* 36, April, Supplement: 101–143.

Green, Venus. 2001. *Race on the Line: Gender, Labor, and Technology in the Bell System, 1880–1980*. Duke University Press.

Greenberg, Dolores. 2000. "Reconstructing Race and Protest: Environmental Justice in New York City." *Environmental History* 5: 223–250.

Greenberg, Jonathan D. 1990. *Staking a Claim: Jake Simmons and the Making of an African-American Oil Dynasty*. Atheneum.

Grim, Valerie. 1994. "The Impact of Mechanized Farming on Black Farm Families in the Rural South: A Study of Farm Life in the Brooks Farm Community, 1940–1970." *Agricultural History* 68, spring: 169–184.

Grim, Valerie. 1995. "The Politics of Inclusion: Black Farmers and the Quest for Agribusiness Participation, 1945–1990s." *Agricultural History* 69, spring: 257–271.

Gropman, Alan. 1998. *The Air Force Integrates, 1945–1964*. Smithsonian Institution Press.

Gubert, Betty Kaplan. 1994. *Invisible Wings: An Annotated Bibliography on Blacks in Aviation, 1916–1993*. Greenwood.

Haber, Louis. 1970. *Black Pioneers of Science and Invention*. Harcourt, Brace.

Hall, Clyde Woodrow. 1973. *Black Vocational, Technical, and Industrial Arts Education: Development and History*. American Technical Society.

Halpern, Rick. 1996. *Meatpackers: An Oral History of Black Packinghouse Workers and Their Struggle for Racial and Economic Equality*. Twayne.

Halpern, Rick. 1997. *Down on the Killing Floor: Black and White Workers in Chicago's Packinghouses, 1904–54*. University of Illinois Press.

Hammonds, Evelynn M. 1997. "New Technologies of Race." In *Processed Lives*, ed. J. Terry and M. Calvert. Routledge.

Harris, Jason T. 1992. "Engineers: 1 in 3 Make It." *Black Enterprise* 23, October: 24.

Harrison, Bennett. 1997. "Bringing High Tech to Low Income People." *Technology Review* 100, April: 64.

Hawkins, Randolph, and Arthur E. Paris. 1997. "Computer Literacy and Computer Use Among College Students: Differences in Black and White." *Journal of Negro Education* 66: 147–158.

Heath, Barbara. 1999. *Hidden Lives: The Archaeology of Slave Life at Thomas Jefferson's Poplar Forest*. University Press of Virginia.

Heinicke, Craig. 1994. "African-American Migration and Urban Labor Skills: 1950 and 1960." *Agricultural History* 68, spring: 185–198.

Henderson, Alexa B. 1990. *Atlanta Life Insurance Company: Guardian of Black Economic Dignity*. University of Alabama Press.

Herman, Bernard L. 1997. "Ode on a Charleston Pitcher." *Natural History* 106, July-August: 63–65.

Higginbotham, Evelyn Brooks, et al., eds. 2001. *The Harvard Guide to African-American History*. Harvard University Press.

Higher Education for Science and Engineering: A Background Paper. 1989. Office of Technology Assessment.

Hill, Susan T. 1992. *Blacks in Undergraduate Science and Engineering Education*. National Science Foundation.

Hoffman, Donna L., and Thomas P. Novak. 1998. "Bridging the Racial Divide on the Internet." *Science* 280, April 17: 390–391.

Hoffman, Steven J. 1993. Behind the Facade: The Constraining Influence of Race, Class and Power on Elites in the City-Building Process, Richmond, Virginia, 1870–1920. Ph.D. dissertation, Carnegie-Mellon University.

Holden, Constance. 1995. "Fewer Black Engineers." *Science* 270, November 24: 1305.

Holley, Donald. 2000. *The Second Great Emancipation: The Mechanical Cotton Picker, Black Migration, and How They Shaped the Modern South*. University of Arkansas Press.

Horowitz, Roger. 1997. *Negro and White, Unite and Fight! A Social History of Industrial Unionism in Meatpacking, 1930–90*. University of Illinois Press.

Horowitz, Roger. 1997. "'Where Men Will Not Work': Gender, Power, Space, and the Sexual Division of Labor in America's Meatpacking Industry, 1890–1990." *Technology and Culture* 38, January: 187–213.

Horton, Carrell, and Jessie Carney Smith, eds. 1990. *Statistical Record of Black America*. Gale Group.

Hunt, Martin K., and Jacqueline E. Hunt. 1999. *The History of Black Business: The Coming of America's Largest African-American-Owned Businesses*. Knowledge Express.

Hurley, Andrew. 1995. *Environmental Inequalities: Class, Race, and Industrial Pollution in Gary, Indiana, 1945–1980*. University of North Carolina Press.

Jackson, Giles B., and D. Webster Davis. 1971. *The Industrial History of the Negro Race of the United States*. Reprint of 1908 edition: Books for Libraries Press.

James, Portia P. 1989. *The Real McCoy: African-American Invention and Innovation, 1619–1930*. Smithsonian Institution Press.

James, Samuel D. K. 1985. *The Impact of Cybernation Technology on Black Automotive Workers in the U.S.* UMI Research Press.

Jaynes, Gerald D. 1986. *Branches Without Roots: Genesis of the Black Working Class in the American South, 1862–1882*. Oxford University Press.

Jaynes, Gerald D. 1988. "Race and Class in Postindustrial Employment." *American Economic Review* 88, May: 356–362.

Jenkins, Edward S. 1984. "Impact of Social Conditions: A Study of the Works of American Black Scientists and Inventors." *Journal of Black Studies* 14, June: 477–491.

Jenkins, Edward. 1991. "Bridging the Two Cultures: American Black Scientists and Inventors." *Journal of Black Studies* 21, March: 313–324.

Johnson, Michael P., and James L. Roark. 1986. *Black Masters: A Free Family of Color in the Old South*. Norton.

Johnson, Robert C. 1993. "Science, Technology and Black Community Development." In *The "Racial" Economy of Science*, ed. S. Harding. Indiana University Press.

Johnson, Whittington Bernard. 1993. *The Promising Years, 1750–1830: The Emergence of Black Labor and Business*. Garland.

Jones, Allen. 1975. "The Role of Tuskegee Institute in the Education of Black Farmers." *Journal of Negro History* 60: 252–267.

Jones, Jacqueline. 1985. *Labor of Love, Labor of Sorrow: Black Women, Work, and the Family from Slavery to the Present*. Basic Books.

Jones, Jacqueline. 1999. *American Work: Four Centuries of Black and White Labor*. Norton.

Jones, James H. 1993. *Bad Blood: The Tuskegee Syphilis Experiment*. Free Press.

Jezierski, John Vincent. 2000. *Enterprising Images: The Goodridge Brothers, African American Photographers, 1847–1922*. Wayne State University Press.

Kahn, Jeffrey, P., Anna C. Mastroianni, and Jeremy Sugarman, eds. 1998. *Beyond Consent: Seeking Justice in Research*. Oxford University Press.

Kaplan, Sidney. 1955. "Jan Earnest Matzeliger and the Making of the Shoe." *Journal of Negro History* 40, January: 8–33.

Karwatka, Dennis. 1991. "Against All Odds." *American Heritage of Invention and Technology* 6, winter: 50–55.

Kenmitzer, Sue. 1988. "Changing America: The New Face of Science and Engineering." *Black Issues in Higher Education*, October 27: 88.

Kenzer, Robert C. 1997. *Enterprising Southerners: Black Economic Success in North Carolina, 1865–1915*. University Press of Virginia.

Klein, Aaron E. 1971. *The Hidden Contributors: Black Scientists and Inventors in America*. Doubleday.

Knox, Robert J. 1993. "Environmental Equity." *Journal of Environmental Health* 55, May: 32–34.

Kolata, Gina. 1991. "In Medical Research Equal Opportunity Doesn't Always Apply." *New York Times*, March 10.

Koos, Greg, and Marcia Young. 1993. "Peter C. Duff: Craft as Biography." *Material Culture* 25, fall: 35–51.

Krieger, Nancy Leys, and Mary Bassett. 1993. "The Health of Black Folk: Disease, Class, and Ideology in Science." In *The "Racial" Economy of Science*, ed. S. Harding. Indiana University Press.

Krueger, Joachim. 1998. "Division on the Internet?" *Science* 281, August 14: 919ff.

Kusmer, Kenneth L. 1978. *A Ghetto Takes Shape: Black Cleveland, 1870–1930*. University of Illinois Press.

Kusmer, Kenneth L., ed. 1991. *Black Communities and Urban Development in America, 1720–1990*. Garland.

Lakwete, Angela. 1994. Cotton Ginning in America, 1780–1890. Ph.D. dissertation, University of Delaware.

Lawrence, Michael A. 1985. "Technology and the Workplace." *The Crisis* 92, April: 22.

Lead Poisoning: Hearing on Impact on Low-Income and Minority Communities. 1992. U.S. Government Printing Office.

Lemann, Nicholas. 1991. *The Promised Land: The Great Black Migration and How It Changed America*. Knopf.

Lerman, Nina. 1997. "'Preparing for the Duties and Practical Business of Life': Technological Knowledge and Social Structure in Mid-19th-Century Philadelphia." *Technology and Culture* 38, January: 31–59.

Lerman, Nina. 1997. "The Uses of Useful Knowledge: Science, Technology, and Social Boundaries in an Industrializing City." *Osiris* 12: 39–59.

Lesseig, Corey T. 2001. *Automobility and Social Change in the South, 1909–1939*. Garland.

Letwin, Daniel. 1998. *The Challenge of Interracial Unionism: Alabama Coal Miners, 1878–1921*. University of North Carolina Press.

Lewis Research Center. 1992. *African American Contributions to Science and Engineering*. Lewis Research Center, National Aeronautics and Space Administration.

Lewis, Ronald L. 1979. *Coal, Iron and Slaves: Industrial Slavery in Maryland and Virginia, 1715–1865*. Greenwood.

Lewis, Ronald L. 1987. *Black Coal Miners in America: Race, Class, and Community Conflict, 1780–1980*. University Press of Kentucky.

Lewis, W. David. 1994. *Sloss Furnaces and the Rise of the Birmingham District: An Industrial Epic*. University of Alabama Press.

Love, Spencie. 1996. *One Blood: The Death and Resurrection of Charles R. Drew*. University of North Carolina Press.

Loving, Neal V. 1994. *Loving's Love: A Black American's Experience in Aviation*. Smithsonian Institution Press.

Lucena, Juan C. 2000. "Making Women and Minorities in Science and Engineering for National Purposes in the United States." *Journal of Women and Minorities in Science and Engineering* 6: 1–32.

Lury, Celia. 1993. *Cultural Rights: Technology, Legality and Personality*. Routledge.

Lynn-Sherow, Bonnie. 1996. *Mechanization, Land Use, and Ownership: Oklahoma in the Early Twentieth Century*. University of Wisconsin Press.

MacDowell, Marsha L., ed. 1997. *African-American Quiltmaking in Michigan*. Michigan State University Press.

MacGregor, Morris J. 2001. *Integration of the Armed Forces, 1940–1965*. Center of Military History, U.S. Army.

Malcolm, Shirley. 1993. "Increasing the Participation of Black Women in Science and Technology." In *The "Racial" Economy of Science*, ed. S. Harding. Indiana University Press.

Mandle, Jay R. 1992. *Not Slave, Not Free: The African-American Economic Experience Since the Civil War*. Duke University Press.

Margo, Robert A. 1990. *Race and Schooling in the South, 1880–1950: An Economic History*. University of Chicago Press.

Marriott, Michel. 1995. "Cybersoul Not Found." *Newsweek* 126, July 31: 62ff.

Marshall, Patricia Phillips. 2001. "The Legendary Thomas Day: Debunking the Popular Mythology of an African-American Craftsman." *North Carolina Historical Review* 78: 32–66.

Massey, Walter E. 1992. "A Success Story Amid Decades of Disappointment." *Science* 258, November 13: 1177–1179.

Matyas, Marsha Lakes, and Shirley M. Malcolm, eds. 1991. *Investing in Human Potential: Science and Engineering at the Crossroads.* American Association for the Advancement of Science.

McCook, Kathleen. 1993. *Toward a Just and Productive Society: An Analysis of the Recommendations of the White Conference on Library and Information Services.* U.S. National Commission on Libraries and Information Science.

McDaniel, George W. 1982. *Hearth and Home: Preserving a People's Culture.* Temple University Press.

McGirr, Lisa. 1995. "Black and White Longshoremen in the IWW: A History of the Philadelphia Marine Transport Workers Industrial Union Local 8." *Labor History* 36, summer: 377–402.

McKissack, Frederick. 1998. "Cyber Ghetto: Blacks are Falling Through the Net." *The Progressive* 2, June: 20–22.

McKiven, Henry M., Jr. 1995. *Iron and Steel: Class, Race, and Community in Birmingham, Alabama, 1875–1920.* University of North Carolina Press.

McLeod, Jonathan W. 1989. *Workers and Workplace Dynamics in Reconstruction-Era Atlanta: A Case Study.* University of California Press.

McMurry, Linda O. 1981. *George Washington Carver: Scientist and Symbol.* Oxford University Press.

Meier, August. 1979. *Black Detroit and the Rise of the UAW.* Oxford University Press.

"Minorities in Science." 1992. *Science* 258, November 13: 1176–1235.

Mitchell, Melvin L. 2001. *The Crisis of the African-American Architect: Conflicting Cultures of Architecture and (Black) Power.* Writers Club Press.

Mohai, Paul. 1990. "Black Environmentalism." *Social Science Quarterly* 71: 744–765.

Muhammad, Tariq K. 1998. "Reading, Writing, and RAM." *Black Enterprise* 28, March: 72ff.

Muhammad, Tariq K., and Cheryl Coward. 1998. "The Black Digerati." *Black Enterprise* 28, March: 49ff.

Murray, Thomas H., Mark A. Rothstein, and Robert F. Murray Jr., eds. 1996. *The Human Genome Project and the Future of Health Care.* Indiana University Press.

Muwakkil, Salim. 1998. "Black America Online." *In These Times*, February 22: 13–14.

Nasaw, David. 1993. *Going Out: The Rise and Fall of Public Amusements.* Basic Books.

Nelson, Bruce. 2001. *Divided We Stand: American Workers and the Struggle for Black Equality.* Princeton University Press.

Newton, James E., and Ronald L. Lewis, eds. 1978. *The Other Slaves: Mechanics, Artisans, and Craftsmen.* G. K. Hall.

Nieman, Donald G. 1994. *African Americans and Non-Agricultural Labor in the South, 1865–1900.* Garland.

Nieman, Donald G., ed. 1994. *From Slavery to Sharecropping: White Land and Black Labor in the Rural South, 1865–1900.* Garland.

Nixon, Ron. 1995. "Winning Water for Keysville." *National Wildlife* 33, August-September: 22–23.

Osur, Alan M. 1976. "Blacks in the Army Air Forces During World War II: The Problem of Race Relations." Office of Air Force History.

Parrish, Deidra Ann. 1997. "The Story Behind the Numbers." *Black Enterprise* 27, January: 35.

Pearson, Kim E. 1983. "Pioneering Black Bell Labs Engineer Still at Work at 71." *The Crisis* 90, April: 40–41.

Peiss, Kathy. 1998. *Hope in a Jar: The Making of America's Beauty Culture*. Metropolitan Books.

Perdue, Robert E. 1975. *Black Laborers and Black Professionals in Early America, 1750–1830*. Vantage.

Peterson, Joyce S. 1979. "Black Automotive Workers in Detroit, 1910–1930." *Journal of Negro History* 64, summer: 177–190.

Phelps, J. Alfred. 1994. *They Had a Dream: The Story of African-American Astronauts*. Presidio.

Powell, William J. 1994. *Black Aviator: The Story of William J. Powell*. Smithsonian Institution Press. Reprint of 1934 publication.

Preston, Howard L. 1979. *Automobile Age Atlanta: The Making of a Southern Metropolis, 1900–1935*. University of Georgia Press.

Putney, Martha S. 1987. *Black Sailors: Afro-American Merchant Seamen and Whalemen Prior to the Civil War*. Greenwood.

Redd, Lawrence. 1988. "Telecommunication, Economics, and Black Families in America." *Journal of Black Studies* 19, September: 111–123.

Reed, Wornie L. 1992. *Health and Medical Care of African-Americans*. William Monroe Trotter Institute, University of Massachusetts.

Reverby, Susan M., ed. 2000. *Tuskegee's Truths: Rethinking the Tuskegee Syphilis Study*. University of North Carolina Press.

Reynolds, Dennis J. 1992. *Citizens Rights and Access to Electronic Information: The 1991 LITA President's Program Presentations and Background Papers*. American Library Association.

Rich, Doris L. 1993. *Queen Bess: Daredevil Aviator*. Smithsonian Institution Press.

Roberts, Dorothy E. 1997. *Killing the Black Body: Race, Reproduction, and the Meaning of Liberty*. Pantheon.

Roberts, J. Timmons, and Melissa M. Toffolon-Weiss. 2001. *Chronicles from the Environmental Justice Frontline*. Cambridge University Press.

Rock, Howard B., Paul A. Gilje, and Robert Asher, eds. 1995. *American Artisans: Crafting Social Identity, 1750–1850*. Johns Hopkins University Press.

Rodriquez, Jessie M. 1989. "The Black Community and the Birth Control Movement." In *Passion and Power*, ed. K. Peiss and C. Simmons. Temple University Press.

Roque, Julie A. 1993. "Environmental Equity: Reducing Risk for All Communities." *Environment* 35, June: 25–28.

Rose, Mark. 1995. *Cities of Light and Heat: Domesticating Gas and Electricity in Urban America*. Pennsylvania State University Press.

Rose, Tricia. 1994. *Black Noise: Rap Music and Black Culture in Contemporary America.* Wesleyan University Press.

Rosengarten, Dale. 1986. *Row Upon Row: Sea Grass Baskets of the South Carolina Low Country.* University of South Carolina.

Salzman, Jack, David Lionel Smith, and Cornel West, eds. 1996. *Encyclopedia of African-American Culture and History.* Macmillan Library Reference.

Sammons, Vivian O., and Denise P. Dempsey. 1989. *Blacks in Science and Related Disciplines.* Library of Congress.

Sandler, Stanley. 1992. *Segregated Skies: All-Black Combat Squadrons of WWII.* Smithsonian Institution Press.

Savage, Barbara Dianne. 1999. *Broadcasting Freedom: Radio, War, and the Politics of Race, 1938–1948.* University of North Carolina Press.

Savitt, Todd L. 2002. *Medicine and Slavery: The Diseases and Health Care of Blacks in Antebellum Virginia.* University of Illinois Press.

Schneider, Janet M., and Bayla Singer, eds. 1995. *Blueprint for Change: The Life and Times of Lewis H. Latimer.* Queens Borough Public Library.

Schon, Donald A., Bish Sanyal, and William J. Mitchell, eds. 1998. *High Technology and Low-Income Communities.* MIT Press.

Schor, Joel, and Cecil Harvey. 1975. *A List of References for the History of Black Americans in Agriculture, 1619–1974.* Agricultural History Center, University of California.

Schwartz, Joel. 1993. *The New York Approach: Robert Moses, Urban Liberals, and Redevelopment of the Inner City.* Ohio State University Press.

Schweninger, Loren. 1989. "Black-Owned Businesses in the South, 1790–1880." *Business History Review* 63: 22–60.

Schweninger, Loren. 1990. *Black Property Owners in the South, 1790–1915.* University of Illinois Press.

Sclove, Richard, and Jeffrey Scheuer. 1994. "The Ghost in the Modem." *Washington Post,* May 29.

Siefert M., G. Gerbner, and J. Fisher. 1989. *The Information Gap: How Computers and Other New Communication Technologies Affect Social Distribution of Power.* Oxford University Press.

Shimkin, David. 1992. "Sex, Race, and Ethnic Diversity: The Media and Minorities." In *State of the Art:*, ed. D. Shimkin et al. St. Martin's.

Simmons-Hodo, Simmona E. 1993. "Silent Builders: A Selected Bibliography on Afro-American Architects." *Bulletin of Bibliography* 50, September: 207–212.

Sims, Calvin. 1992. "From Inner-City L.A. to Yale Engineering." *Science* 258, November 13: 1232.

Singleton, Theresa A. 1995. "The Archaeology of Slavery in North America." *Annual Reviews of Anthropology* 24: 119–140.

Singleton, Theresa A., ed. 1997. *The Archaeology of Slavery and Plantation Life.* Academic Press.

Singleton, Theresa A., ed. 1999. *"I, Too, Am America": Archaeological Studies of African-American Life*. University Press of Virginia.

Smith, E. Valerie. 1993. "The Black Corps of Engineers and the Construction of the Alaska Highway." *Negro History Bulletin* 51, December: 22.

Smith, Jessie Carney, and Carrell Peterson Horton, eds. 1994. *Historical Statistics of Black America: Agriculture to Labor and Employment*. Gale Group.

Smith, Julia Floyd. 1992. *Slavery and Rice Culture in Low Country Georgia, 1750–1860*. University of Tennessee Press.

Smith, Susan L. 1995. *Sick and Tired of Being Sick and Tired: Black Women's Health Activism in America, 1890–1950*. University of Pennsylvania Press.

Snider, Jill D. 1995. Flying to Freedom: African-American Visions of Aviation, 1910–1927. Ph.D. dissertation, University of North Carolina.

Spivey, Donald. 1978. *Schooling for the New Slavery: Black Industrial Education, 1868–1915*. Greenwood.

Stanley, Autumn. 1983. "From Africa to America: Black Women Inventors." In *The Technological Woman*, ed. J. Zimmerman.

Stanley, Autumn. 1995. *Mothers and Daughters of Invention: Notes for a Revised History of Technology*. Rutgers University Press.

Starobin, Robert S. 1970. *Industrial Slavery In the Old South*. Oxford University Press.

Steward, Jeffrey C. 1996. *1001 Things Everyone Should Know about African American History*. Doubleday.

Stewart, James B. 1986. "STS and Black Studies: Partnership for Progress in the 21st Century." *Bulletin of Science, Technology, and Society* 6: 315–318.

Stone, Richard. 1993. "Industrial Efforts: Plenty of Jobs, Little Minority Support in Biotech." *Science* 262, November 12: 1127.

Stroud, Ellen. 1999. "Troubled Waters in Ecotopia: Environmental Racism in Portland, Oregon." *Radical History Review* 74: 65–95.

Symposium on Minorities and Women in Science and Technology: A Report. 1982. U.S. Government Printing Office.

Tapper, Melbourne. 1999. *In the Blood: Sickle Cell Anemia and the Politics of Race*. University of Pennsylvania Press.

Taylor, Dorceta E. 1989. "Blacks and the Environment: Toward an Explanation of the Concern and Action Gap between Blacks and Whites." *Environment and Behavior* 21, March: 175–205.

Taylor, Henry Louis, Jr., ed. 1993. *Race and the City: Work, Community, and Protest in Cincinnati, 1820–1970*. University of Illinois Press.

Taylor, Henry Louis, Jr. 1986. "On Slavery's Fringe: City-Building and Black Community Development in Cincinnati, 1800–1850." *Ohio History* 95, winter-spring: 5–33.

Taylor, Henry Louis, and Walter Hill, eds. 2000. *Historical Roots of the Urban Crisis: Blacks in the Industrial City, 1900–1950*. Garland.

The African American Mosaic: A Library of Congress Resource Guide for the Study of Black History and Culture. 1994. Library of Congress.

"The Push for Blacks in Technology." 1989. *Black Enterprise* 19, May: 61.

Thomas, Bettye C. 1974. "A Nineteenth-Century Black Operated Shipyard, 1866–1884." *Journal of Negro History* 59: 1–12.

Thomas, Valerie L. 1989. "Black Women Engineers and Technologists." *Sage* 6, fall: 24–32.

Tobin, Jacqueline L., and Raymond G. Dobard. 1999. *Hidden in Plain View: A Secret Story of Quilts and the Underground Railroad*. Doubleday.

Toxic Wastes and Race in the United States. 1987. United Church of Christ Commission for Social Justice.

Trent, William, and John Hill. 1994. "The Contributions of Historically Black Colleges and Universities to the Production of African-American Scientists and Engineers." In *Who Will Do Science?* ed. W. Pearson and A. Fechter. Johns Hopkins University Press.

Trotter, Joe W. 1990. *Coal, Class, and Color: Blacks in Southern West Virginia, 1915–32*. University of Illinois Press.

Trotter, Joe W., and Earl Lewis, eds. 1996. *African-Americans in the Industrial Age: A Documentary History, 1915–1945*. Northeastern University Press.

Ulrich, Laurel Thatcher. 2001. *The Age of Homespun: Objects and Stories in the Creation of an American Myth*. Knopf.

United States Congress, House Committee on Energy and Commerce. 1990. *Health Status and Needs of Minorities in the 1990s: Hearing*. U.S. Government Printing Office.

United States Congress, Senate Committee on Labor and Human Resources. 1993. *The Disadvantaged Minority Health Improvement Act: Hearing and Report*. U.S. Government Printing Office.

United States Department of Health and Human Services. 1991. *Health Status of Minorities and Low-Income Groups*. U.S. Government Printing Office.

United States Department of Health and Human Services, Office of Minority Health. 1993. *Toward Equality of Well-Being: Strategies for Improving Minority Health*. U.S. Government Printing Office.

United States Task Force on Women, Minorities, and the Handicapped in Science and Technology. 1989. *Changing America: The New Face of Science and Engineering, Final Report*.

Urban Habitat Program of Earth Island Institute. *Race, Poverty, and the Environment*. Quarterly newsletter.

Velez-Rodriguez, Argelia. 1986. "Technological Literacy and Ethnic Minority Students." *Bulletin of Science, Technology, and Society* 6: 311–314.

Vlach, John Michael. 1978. *The Afro-American Tradition in Decorative Arts*. Cleveland Museum of Art.

Vlach, John Michael. 1991. *By the Work of Their Hands: Studies in Afro-American Folklife*. University Press of Virginia.

Vlach, John Michael. 1993. *Back of the Big House: The Architecture of Plantation Slavery*. University of North Carolina Press.

Wahlman, Maude Southwell. 2001. *Signs and Symbols: African Images in African American Quilts*, second edition. Tinwood.

Wailoo, Keith. 1997. *Drawing Blood: Technology and Disease Identity in Twentieth-Century America*. Johns Hopkins University Press.

Wailoo, Keith. 2001. *Dying in the City of the Blues: Sickle Cell Anemia and the Politics of Race and Health*. University of North Carolina Press.

Waks, Leonard J. 1991. "Technological Literacy for the New Majority." In *Science Education in the United States: Issues*, ed. S. Majumdar et al. Pennsylvania Academy of Science.

Walker, Ballus. 1991. "Environmental Health and African-Americans." *American Journal of Public Health* 81, November: 1395–1398.

Walker, Juliet E. K. 1998. *The History of Black Business In America: Capitalism, Race, Entrepreneurship*. Macmillan Library Reference.

Walker, Juliet E. K., ed. 1999. *Encyclopedia of African-American Business History*. Greenwood.

Waller, Gregory A. 1995. *Main Street Amusements: Movies and Commercial Entertainment in a Southern City, 1896–1930*. Smithsonian Institution Press.

Walsh, R. Taylor. 1994. *The National Information Infrastructure and The Recommendations of the 1991 White House Conference on Library and Information Services*. U.S. National Commission on Libraries and Information Science.

Winslow, Calvin, ed. 1998. *Waterfront Workers: New Perspectives on Race and Class*. University of Illinois Press.

Winslow, E., ed. 1974. *Black Americans in Science and Engineering: Contributions of Past and Present*. Afro-American Press for General Electric.

Weber, Paul R. 1988. "Antoine Goes Home." *Civil Engineering* 58, March: 6.

Webster, Raymond B. 2000. *African American Firsts in Science and Technology*. Gale Group.

Weitzner, Daniel J. 1994. "Building Open Platforms: Public Policy for The Information Age." *Telecommunications* 28, January: 79–82.

Wexler, Mark. 1993. "Sweet Tradition." *National Wildlife* 31, APril-May: 38–41.

Wharton, David E. 1992. *A Struggle Worthy of Note: The Engineering and Technological Education of Black Americans*. Greenwood.

White, Patricia E. 1992. *Women and Minorities in Science and Engineering: An Update*. National Science Foundation.

White, Shane, and Graham White. 1995. "Slave Hair and African-American Culture in the Eighteenth and Nineteenth Centuries." *Journal of Southern History* 61, February: 45–76.

White, Shane, and Graham White. 1998. *"Stylin': African American Expressive Culture from Its Beginnings to the Zoot Suit*. Cornell University Press.

Whitmore, Kay R. 1988. "The Growing Need for Minority Engineers." *Chemical Engineering Progress* 84, January: 57–59.

Whitten, David O. 1981. *Andrew Durnford: A Black Sugar Planter in Antebellum Louisiana*. Northwestern State University Press.

Wilkerson, Isabel. 1991. "Blacks Assail Ethics in Medical Testing." *New York Times*, June 3.

Wilkie, Laurie. 2000. *Creating Freedom: Material Culture and African American Identity at Oakley Plantation, Louisiana, 1840–1950*. Louisiana State University Press.

Wilkinson, R. Keith. 1990. *Science and Engineering Personnel: A National Overview*. National Science Foundation.

Williams, Clarence G. 2001. *Technology and the Dream: Reflections on the Black Experience at MIT, 1941–1999*. MIT Press.

Willis, Deborah, and Robin D. G. Kelley. 2000. *Reflections in Black: A History of Black Photographers, 1840 to the Present*. Norton.

Winner, Langdon. 1986. "Do Artifacts Have Politics?" In Winner, *The Whale and the Reactor*. University of Chicago Press.

Worthy, Ward. 1990. "Minorities Speak Out on Science Education Needs." *Chemical and Engineering News*, October 1: 28–30.

Wresch, William. 1996. *Disconnected: Haves and Have-Nots in the Information Age*. Rutgers University Press.

Wright, Gavin. 1986. *Old South, New South: Revolutions in the Southern Economy Since the Civil War*. Basic Books.

Yancy, Dorothy Cowser. 1984. "The Stuart Double Plow and Double Scraper: The Invention of a Slave." *Journal of Negro History* 69, winter: 37–51.

Zafar, Rafia. 1999. "The Signifying Dish: Autobiography and History in Two Black Women's Cookbooks." *Feminist Studies* 25: 449–469.

Notes

1. As a general introduction to this field, see John M. Staudenmaier, *Technology's Storytellers: Reweaving the Human Fabric* (Society for the History of Technology and MIT Press, 1985).

2. Alan I. Marcus and Howard P. Segal, *Technology in America: A Brief History* (Harcourt Brace, 1999); Ruth Schwartz Cowan, *A Social History of American Technology* (Oxford University Press, 1997); Carroll Pursell, *The Machine in America: A Social History of Technology* (Johns Hopkins University Press, 1995); Gary Cross and Rick Szostak, *Technology and American Society: A History* (Prentice-Hall, 1995).

3. For discussions of the history of women, gender, and technology, see Judith A. McGaw, "Women and the History of American Technology," *Signs* 7 (1982): 798–828; Judith A. McGaw, "No Passive Victims, No Separate Spheres: A Feminist Perspective on Technology's History," in *In Context*, ed. S. Cutcliffe and R. Post (Lehigh University Press, 1989); Ruth Schwartz Cowan, "From Virginia Dare to Virginia Slims: Women and Technology in American Life," *Technology and Culture* 20 (1979): 51–63. For one of the most recent analyses, see Nina Lerman, Arwen Palmer Mohun, and Ruth Oldenziel, "The Shoulders We Stand on and the View from Here: Historiography and Directions for Research," *Technology and Culture* 38 (1997): 9–30 (see also the other six articles in that issue). As just a few examples of specific studies on gender and technology, see Ruth Schwartz Cowan, *More Work for Mother: The Ironies of Household Technology from the Open Hearth to the Microwave* (Basic Books, 1983); Virginia Scharff, *Taking the Wheel: Women and the Coming of the Motor Age* (Free Press, 1991); Jane Farrell-Beck and Colleen Gau, *Uplift: The Bra in America* (University of Pennsylvania Press, 2001); Judy Wajcman, *Feminism Confronts Technology* (Pennsylvania State University Press, 1991).

4. For more on politics and the social construction of technology, see Michael Adas, *Machines as the Measure of Men: Science, Technology, and Ideologies of Western Dominance* (Cornell University Press, 1989); Wiebe E. Bijker, Thomas P. Hughes, and Trevor Pinch, eds., *The Social Construction of Technological Systems: New Directions in the Sociology and History of Technology* (MIT Press, 1987); Rudi Volti, *Society and Technological Change* (Worth, 2001); Donald MacKenzie and Judy Wajcman, eds., *The Social Shaping of Technology* (Open University Press, 1999).

5. To take just a few illustrative and general examples, see John Hope Franklin and Alfred A. Moss Jr., *From Slavery to Freedom: A History of African-Americans* (McGraw-Hill, 1994); Darlene Clark Hine, ed., *Black Women in United States History* (Carlson, 1990); Genevieve Fabre and Robert O'Meally, eds., *History and Memory in African-American Culture* (Oxford University Press, 1994). For a good college-level essay reader, see William R. Scott and William G. Shade, eds., *Upon These Shores: Themes in the African-American Experience, 1600 to the Present* (Routledge, 1999).

6. As starting points, see Kenneth R. Manning, *Black Apollo of Science: The Life of Ernest Everett Just* (Oxford University Press, 1983); Sandra Harding, ed., *The "Racial" Economy of Science: Toward a Democratic Future* (Indiana University Press, 1993).

7. For a start, see Darlene Clark Hine, *Black Women in White: Racial Conflict and Cooperation in the Nursing Profession, 1890–1950* (Indiana University Press, 1989), especially the chapter "Co-Laborers in the Work of the Lord: Nineteenth-Century Black Women Physicians."

8. For a valuable discussion about expanding a narrow conception of "technology," see Judith A. McGaw, ed., *Early American Technology: Making and Doing Things from the Colonial Era to 1850* (University of North Carolina Press, 1994).

9. For a general discussion of the history of invention, patenting, and the diffusion of innovation, see Carolyn Cooper, "Making Inventions Patent," *Technology and Culture* 32 (1991), October: 837–845, and the following nine articles in that issue. See also biographies and books on specific inventors, such as Paul Israel, *Edison: A Life of Invention* (Wiley, 1998); Robert Friedel and Paul Israel, *Edison's Electric Light: Biography of an Invention* (Rutgers University Press, 1986); Neil Baldwin, *Edison: Inventing the Century* (Hyperion, 1995); Peter Jakab, *Visions of a Flying Machine: The Wright Brothers and the Process of Invention* (Smithsonian Institution Press, 1990).

10. For an overview of the history of African-American farming and relationship to the land, see Charlene Gilbert and Quinn Eli, *Homecoming: The Story of African-American Farmers* (Beacon, 2000).

11. For background in American agricultural history, see Douglas Hurt, *American Agriculture* (Iowa State University Press, 1994); Willard W. Cochrane, *The Development of American Agriculture: A Historical Analysis* (University of Minnesota Press, 1993).

12. Readers can often find discussions of material culture and the technologies of everyday living incorporated into books which discuss slave life more generally; for instance, there is substantive material on plantation slave housing, diet, dress, skilled work, and artisanship in Philip Morgan, *Slave Counterpoint: Black Culture in the Eighteenth-Century Chesapeake and Lowcountry* (University of North Carolina Press, 1998).

13. Ruth Schwartz Cowan, "The Consumption Junction: A Proposal for Research Strategies in the Sociology of Technology," in *The Social Construction of Technological Systems:*, ed. W.

Bijker et al. (MIT Press, 1987). By contrast, there is a growing literature on the history of gender, consumerism, and technology, especially referring to the embrace of domestic appliances. See Joy Parr, "What Makes Washday Less Blue? Gender, Nation, and Technology Choice in Postwar Canada," *Technology and Culture* 38 (1997): 153–186; Joy Parr, "Shopping for a Good Stove: A Parable about Gender, Design, and the Market," in *His and Hers*, ed. R. Horowitz and A. Mohun (University Press of Virginia, 1998); Katherine Jellison, "'Let Your Corn Stalks Buy You a Maytag': Prescriptive Literature and Domestic Consumption in Rural Iowa," *Palimpsest* 69 (1988): 132–139.

14. As a starting point, see Philip Scranton, "None-Too-Porous Boundaries: Labor History and the History of Technology," *Technology and Culture* 29 (1988), October: 722–743, and the other five articles in the same issue.

15. For a good overview of African-American life in New York City (dramatized with a number of excellent photographs), see Howard Dodson et al., *The Black New Yorkers: The Schomberg Illustrated Chronology* (Wiley, 1999).

16. Similarly, there is an extensive literature on the role of African-Americans in creating and consuming forms of popular communication, including radio and television. Much of this work falls more in the study of popular culture, sociology, or psychology, rather than history of technology. Still, anyone with an interest in how racialized messages have been conveyed in modern media may want to consult William Barlow, *Voice Over: The Making of Black Radio* (Temple University Press, 1998); Herman Gray, *Watching Race: Television and the Struggle for the Sign of Blackness* (University of Minnesota Press, 1997); and Donald Bogle, *Primetime Blues: African Americans on Network Television* (Farrar, Straus & Giroux, 2001).

17. For example, see Bernard Nalty, *Strength for the Fight: A History of Black Americans in the Military* (Free Press, 1986); Gerald Astor, *The Right to Fight: A History of African Americans in the Military* (Presidio, 1998); Gail Lumet Buckley, *American Patriots: The Story of Blacks in the Military from the Revolution to Desert Storm* (Random House, 2001). See also Lenwood Davis and George Hill, eds., *Blacks in the American Armed Forces, 1776–1983: A Bibliography* (Greenwood, 1985).

18. As an introduction to the history and sociology of medicine in America, see Paul Starr, *The Social Transformation of Medicine in America* (Basic Books, 1982); Susan M. Reverby and David Rosner, eds., *Health Care in America: Essays in Social History* (Temple University Press, 1979). On the history of biomedical technology, see Ruth Schwartz Cowan, "Descartes's Legacy," *Technology and Culture* 34 (1993), October: 721–728 and the seven articles following in that issue. See also Stanley Joel Reiser, *Medicine and the Reign of Technology* (Cambridge University Press, 1978). On current policy issues surrounding the development of high-tech medicine and access to expensive care, see Institute for the Future, *Health and Health Care 2010* (Jossey-Bass, 2000); David Ellis, ed., *Technology and the Future of Health Care* (Jossey-Bass, 2000).

19. As background on the history of computers and information-age technology in recent decades, see Steven Lubar, *Infoculture* (Houghton Mifflin, 1993); Paul Ceruzzi, *A History of Modern Computing* (MIT Press, 1998). Books on gender patterns in computer use can prove a good entry point for analysis about the racial digital divide, suggesting some useful lines of questioning and analysis. See Lynn Cherny and Elizabeth Reba Weise, eds., *Wired Women: Gender and New Realities in Cyberspace* (Seal, 1996); Justine Cassell and Henry Jenkins, eds., *From Barbie to Mortal Kombat: Gender and Computer Games* (MIT Press, 1998); Roberta Furger, *Does Jane Compute? Preserving Our Daughters' Place in the Cyber Revolution* (Warner, 1998).

20. For another introduction to the history of race and education, see Leo McGee and Harvey G. Neufeldt, *Education of the Black Adult in the United States: An Annotated Bibliography* (Greenwood, 1985).

21. For background on the history of engineering education, see David Noble, *America by Design* (Knopf, 1977); George Emmerson, *Engineering Education: A Social History* (Crane, Russak, 1973); Terry S. Reynolds, ed., *The Engineer in America* (University of Chicago Press, 1991); Bruce Seely, "Research, Engineering, and Science in American Engineering Colleges, 1900–1960," *Technology and Culture* 34 (1993), April: 344–345. Exploring the link between gender, masculinity, and engineering education may offer some insights on the topic of race. See Ruth Oldenziel, *Making Technology Masculine: Men, Women, and Modern Machines in America, 1870–1945* (Amsterdam University Press, 1999); Cynthia Cockburn, *Machinery of Dominance: Women, Men, and Technical Know-How* (Pluto, 1985); and Amy Bix, "Feminism Where Men Predominate: The History of Women's Science and Engineering Education at MIT," *Women's Studies Quarterly* 28 (2000), spring-summer, pp. 24–45; Amy Bix, "'Engineeresses' Invade Campus: Four Decades of Debate over Technical Coeducation," *IEEE Technology and Society Magazine* 19 (2000), spring: 20–26.

22. Thanks to the following for their help in suggesting bibliographical material: Rosie L. Albritton, Jonathan Coopersmith, Susan E. Cozzens, F. Elaine DeLancy, John Douard, Deborah G. Douglas, Mark S. Frankel, Steven J. Hoffman, Laura Kramer, Judith McGaw, Ed Morman, Richard Sclove, Bruce Sinclair, Bayla Singer, R. Samuel Winningham.

Contributors

Amy Sue Bix is an associate professor in the History Department at Iowa State University and a faculty member of Iowa State's Program in the History of Technology and Science. Her book *Inventing Ourselves Out of Jobs?: America's Debate Over Technological Unemployment, 1921–1981* was published in 2000. She has also written on the history of breast cancer and AIDS research, on the history of eugenics, the history of ISU's home economics program, and on post-World War II physics and engineering. She is currently finishing a book entitled *Engineering Education for American Women: An Intellectual, Institutional, and Social History*.

Lonnie Bunch is president of the Chicago Historical Society. He came to Chicago from the Smithsonian's National Museum of American History, where he was associate director for curatorial affairs and, in his last year there, co-curator of an exhibit entitled *The American Presidency: A Glorious Burden*. In 2002 he was appointed a member of the Committee for the Preservation of the White House.

Judith Carney is professor of geography at the University of California, Los Angeles. Her specialty if the historical ecology of food and agricultural systems in West Africa and Latin America, as well as their linkage to the African diaspora. She has written widely on this theme, including *Black Rice: The African Origins of Rice Cultivation in the Americas* (2001).

Kathleen Franz is assistant professor of history at the University of North Carolina in Greensboro, where she also coordinates the public history program. Her forthcoming book is entitled *Tinkering with Automobility: Consumers and Technological Authority in the Twentieth Century*. She is the 2002 recipient of the Hindle Prize in the history of invention given by the Society for the History of Technology. In her spare time, she works as a guest curator and as a sailboat varnishing specialist.

Barbara Garrity-Blake taught at East Carolina University for nine years and is now a policy maker on the North Carolina Marine Fisheries Commission. She is currently undertaking an ethnohistorical study of the Outer Banks. She is author of *The Fish Factory: Work and Meaning for Black and White Fishermen in the American Menhaden Fishery* (1994), and *Fish House Opera* (with Susan West, in press).

Rebecca Herzig teaches courses on race, gender, science, and technology at Bates College. She is working on her first book, *Suffering for Science: Faith, Reason, and Sacrifice in Fin-de-Siècle American Life*.

Portia James is curator and historian at the Anacostia Museum of the Smithsonian Institution. She was responsible for an wide-ranging exhibition about African-American invention and inventiveness whose catalog was published as *The Real McCoy: African American Invention and Innovation, 1619–1930* (1989). Other exhibits she has developed include *Down Through the Years*, on the Anacostia Museum's own history and collections, and *Black Mosaic: Color, Race, and Ethnicity Among Black Immigrants to Washington, D.C.* She is the author of a forthcoming book titled *Black Washingtonians*, and she is currently working on a history of African-American celebrations and holidays.

Nina Lerman is associate professor of history at Whitman College in Walla Walla, Washington. With Arwen Mohun and Ruth Oldenziel she edited *Gender and Technology: A Reader* (2003). Her explorations in the essay included in this volume grew out of her ongoing study of children's technical education and changing definitions of social boundaries in nineteenth-century Philadelphia.

Bruce Sinclair retired in 1996 from Georgia Tech, where he was the Melvin Kranzberg Professor of the History of Technology. Previously, he had taught for a number of years in the Institute for the History and Philosophy of Science and Technology at the University of Toronto. Long active in the Society for the History of Technology, he served as its president and was awarded its da Vinci medal. He has written on the history of technical education and on engineering education, and is currently working on a book about the engineers involved in the celebrated controversy over the damming of the Hetch Hetchy Valley in Yosemite National Park.

Amy Slaton is an associate professor of history at Drexel University. Her early work focused on the history of materials standards and building construction and includes the book *Reinforced Concrete and the Modernization of American Building, 1900–1930* (2001ᵃ). She is currently investigating the historical exclusion of African-Americans from technical occupations in the United States through a study of engineering education since 1945.

Index